T0177652

PHYSICS OF SPIN-ORBIT-COUPLED OXIDES

Physics of Spin-Orbit-Coupled Oxides

Gang Cao
University of Colorado at Boulder

Lance E. DeLong
University of Kentucky

OXFORD
UNIVERSITY PRESS

OXFORD

UNIVERSITY PRESS

Great Clarendon Street, Oxford, OX2 6DP,
United Kingdom

Oxford University Press is a department of the University of Oxford.
It furthers the University's objective of excellence in research, scholarship,
and education by publishing worldwide. Oxford is a registered trade mark of
Oxford University Press in the UK and in certain other countries

First Edition published in 2021

Impression: 1

Published in the United States of America by Oxford University Press
198 Madison Avenue, New York, NY 10016, United States of America

British Library Cataloguing in Publication Data

Data available

Library of Congress Control Number: 2020952113

ISBN 978–0–19–960202–5

DOI: 10.1093/oso/9780199602025.001.0001

Printed and bound by
CPI Group (UK) Ltd, Croydon, CR0 4YY

Gang Cao dedicates this book to his mother, Xiaomei Xiao,
who has shaped his life

Preface

Transition metal oxides are constantly surprising us with exotic phenomena and challenging existing theoretical models. The advancement of condensed matter and materials physics has been largely dependent on the arduous path toward understanding these materials. The surprising insulating behavior inherent in binary transition metal oxides such as NiO reported in 1937 by De Boer and Verwey led to the realization of the importance of electron-electron correlations first proposed by Peierls and Mott; the high-temperature superconductivity in ternary transition metal oxides $(La_{1-x}Ba_x)_2CuO_4$ discovered in 1986 by Bednorz and Muller violates the Bardeen-Cooper-Schrieffer theory that otherwise perfectly describes conventional superconductivity. Establishing an adequate mechanism driving the high-temperature superconductivity in the cuprates has remained a profound intellectual challenge to this day. Prior to 2010, the overwhelming balance of interest was justifiably devoted to the superconducting cuprates, colossal magnetoresistive manganites and other $3d$-transition metal oxides. In 1994, the discovery of an exotic superconducting state in Sr_2RuO_4 shifted some interest toward ruthenates, and in 2008, the realization that a novel variant of the Mott state was at play in Sr_2IrO_4 provided impetus for a burgeoning group of studies of the influence of strong spin-orbit interactions in "heavy" ($4d$- and $5d$-) transition metal oxides. These materials are now among the most current and intriguing topics in contemporary condensed matter and materials physics.

In the early 1990s, one of us (GC) became interested in new materials, particularly those containing no $3d$-transition metals, and started exploring ruthenates, rhodates, and iridates in search of novel materials and phenomena. This research effort has been intensified and extended over the last two decades. We, the authors, have closely collaborated since 2007. This book is in part based on our work; it focuses on recent experimental and, to a lesser extent, theoretical evidence that the physical and structural properties of these oxides are decisively influenced by strong spin-orbit interactions that compete or collaborate with comparable Coulomb interactions. A hallmark of these materials is the extreme susceptibility to lattice distortions or disorder because of the strong spin-orbit interactions, which lead to unusual ground states and other exotic phenomena unique to this class of materials.

This book consists of six chapters. Chapter 1 reviews underlying electronic interactions and structural or orbital effects that are most relevant to unusual behaviors of $4d$- and $5d$-transition metal oxides. Chapters 2–5 form the central part of the book, in which the distinct physical properties of the $4d$- and $5d$-transition metal oxides that have been discovered or studied in recent years are discussed. These chapters describe basic structural, transport, thermodynamic and magnetic properties of ruthenates and iridates as functions of temperature, pressure, magnetic field and electrical current, with a focus on experimental results and empirical trends. Chapter 6 introduces a few methods of single-crystal synthesis techniques, including a newly developed field-altering technology, that are most suitable for the $4d$- and $5d$-transition metal oxides. This chapter is intended to help fill an existing hiatus in the literature describing relevant synthesis techniques for $4d$- and $5d$-transition metal oxides.

The aim of this book is to provide an introduction to those who are interested in this class of materials and wish to have an accessible survey of the field. The reader is cautioned that this book reflects the interests of the authors. We do not intend a complete review because that is nearly impossible given the rapidly developing field and its diverse nature. There are a number of excellent reviews and books, which are listed as Further Reading at the end of each chapter.

I (GC) would like to thank my coauthor, Lance DeLong, for his critical reading and revisions of the book draft.

I am very grateful to the *National Science Foundation* for its support over the last two decades, which has made this book possible.

We both would like to extend our deep gratitude to Sonke Adlung and his colleagues at Oxford University Press for the continued encouragement and support during the long course of writing this book.

I am profoundly indebted to the then Director of the National High Magnetic Field Laboratory, Jack E. Crow, who was both my PhD advisor and postdoc mentor; during the heyday of the cuprates and mangnites in the 1990s, it was his strong support and encouragement that made the early exploratory studies of $4d$- and $5d$-transition metal oxides possible. I am grateful to Pedro Schlottmann for his theoretical guidance and our longstanding collaboration over the last two decades, which has proven to be crucial to the investigations of these materials. I also wish to express my gratitude to Feng Ye. Over the last ten years of our productive collaboration, his expertise in neutron diffraction has helped deepen the understanding of an array of new materials with unusual structural and magnetic behaviors. I am also grateful to Lance Cooper for our longtime, productive collaboration; his expertise in Raman scattering has helped gain much-needed insights into the physics of these materials. I have immensely enjoyed conversations and collaborations with Peter Riseborough whose insights into physics are always intellectually stimulating. I thank my colleague friends Peter Baker, Stephen Blundell, Yue Cao, Songxue Chi, Mark Dean, Daniel Dessau, Yang Ding, Vladimir Dobrosavljevic, Nuh Gedik, John Goodenough, Daniel Haskel, Michael Hermele, David Hsieh, Changqing Jin, Jiangping Hu, Ribhu Kaul, Daniel Khomskii, Young-June Kim, Itamar Kimchi, Minhyea Lee, Ganpathy Murthy, Tae-Won Noh, Natalia Perkins, Dmitry Reznik, Thorsten Schmitt, Ambrose Seo, Sergey Streltsov, Wenhai Song, Yuping Sun, Mingliang Tian, Darius Torkinsky, Maxim Tsoi, Xiangang Wan, Yan Xin, Zhaorong Yang, Liuyan Zhao, Jianshi Zhou and Yonghui Zhou. Collaborations with them have led to important results presented here. I would also like to express my deep gratitude to my former and current students and postdocs, particularly Hengdi Zhao, Jasminka Terzic, Hao Zheng, Vinobalan Durairaj, Shalinee Chikara, Shujuan Yuan, Saicharan Aswartham, Bing Hu, Jinchen Wang, Tongfei Qi and Oleksandr Korneta whose contributions constitute the most recent and important experimental results presented in the book. I am especially thankful to Hengdi Zhao and Bing Hu for their critical reading of the book proof, which has greatly improved the book. I am indebted to Leon Balents, Bernd Buchner, Gang Chen,

Sang-Wook Cheong, Elbio Dagotto, Xi Dai, Xia Hong, George Jackeli, Rongying Jin, Hae Young Kee, Yong Bin Kim, Giniyat Khaliullin, Junming Liu, David Mandrus, Igor Mazin, John Mitchell, Janice Musfeldt, Ward Plummer, Charles Rogers, Tanusri Saha-Dasgupta, Hide Takagi, Nandini Trivedi, Roser Valenti, Jeroen van den Brink, Nanling Wang, William Witczak-Krempa and Jiandi Zhang for stimulating discussions in recent years.

Most of all, my heartfelt thanks go to my parents, Baigui Cao and Xiaomei Xiao, my wife, Qi Zhou, and my sons, Eric, Vincent, and Tristan, and my sister, Yang Cao, for their support, inspiration, patience, and love.

Gang Cao

Boulder, Colorado
September 2020

Contents

Part 1

Fundamental Principles

Chapter 1

Introduction

1.1 Overview

Transition metal oxides are of great significance and importance both fundamentally and technologically. The emergence of novel quantum states stemming from a delicate interplay between fundamental energies, such as the on-site Coulomb repulsion, spin-orbit interactions (SOI), crystal fields, and Hund's rule coupling, is a key, unique characteristic of these materials.

The transition metals comprise the majority of stable, useful elements in the Periodic Table. Their physical and chemical properties are dominated by their outermost d-electrons, and they readily form compounds with oxygen and sulfur, selenium and tellurium. However, oxides are more abundant than sulfides, selenides or tellurides, in part because oxygen is the most abundant element on Earth, making up 46.6% of the mass of the Earth's crust. It is more than 1000 times as abundant as sulfur, and still more abundant than selenium or tellurium. For example, there are more than 10,000 ternary oxides, and the corresponding number for sulfides, selenides and tellurides is at least one order of magnitude smaller [1]. The most common crystal structures for ternary oxides are the perovskites and pyrochlores. Moreover, oxygen has the strongest electronegativity among the four VIA elements, and therefore transition metal oxides tend to be generally more magnetic, more ionic and thus less conductive, compared to their sulfide, selenide and telluride counterparts.

Transition metal oxides have been the focus of enormous activity within both the applied and basic science communities over the last four decades. However, between the 19th and the late 20th Centuries, the overwhelming balance of interest was devoted to 3d-elements and their binary compounds, chiefly because these materials are more accessible, and their strong magnetic, electronic and elastic properties found many important applications in the Industrial Revolution and the modern "High Tech" era. In particular, John Goodenough pioneered studies of transition metal oxides such as manganites and cobaltates in the 1950s and 1960s; his studies emphasized metal-insulator transitions and magnetic properties of these 3d-materials. Certain semi-empirical rules developed at that time form the basis of the concept of superexchange interactions, which are at the heart of cooperative magnetic phenomena in transition metal oxides [2,3].

Physics of Spin-Orbit-Coupled Oxides. Gang Cao and Lance E. DeLong, Oxford University Press (2021). © Gang Cao and Lance E. DeLong.
DOI: 10.1093/oso/9780199602025.003.0001

Table 1.1 *Transition Metals with Common Oxidation States and Ionic Radii*

Z	22	23	24	25	26	27	28	29	30
3d	Ti	V	Cr	Mn	Fe	Co	Ni	Cu	Zn
Oxidation state	+2,+3,+4	+2,+3,+4	+2,+3,+4	+2,+3,+7	+2,+3,+4	+2,+3,+4	+2,+3,+4	+1,+2,+3	+2
Ionic radius (pm)	86,67,60.5	79,64,58	80,61.5,55	83,64.5,53	78,64.5,58.5	74.5,61,53	69,60,48	77,73,54	74
	40	41	42	43	44	45	46	47	48
4d	Zr	Nb	Mo	Tc	Ru	Rh	Pd	Ag	Cd
	+4	+3,+4,+5	+3,+4,+5	+4,+5,+7	+3,+4,+5	+3,+4,+5	+2,+3,+4	+1,+2,+3	+2
	72	72,68,64	69,65,61	64.5,60,56	68,62,56.5	66.5,60,55	86,76,61.5	115,94,75	95
	72	73	74	75	76	77	78	79	80
5d	Hf	Ta	W	Re	Os	Ir	Pt	Au	Hg
	+4	+3,+4,+5	+4,+5,+6	+4,+5,+6	+4,+5,+6	+3,+4,+5	+2,+4,+5	+1,+3,+5	+1,+2
	72	72,68,64	66,62,60	63,58,55	63,57.5,54.5	68,62.5,57	80,62.5,57	137,62,60	119,102

From the 1950s to the 1980s, the robust superconducting properties of several 4d-based Nb intermetallic compounds such as Nb_3Sn and NbTi shifted some attention toward the 4d-elements and compounds. Remarkably, Nb_3Sn remains the most widely used superconductor to this day, despite thousands of known superconducting materials. A sea change occurred with the discovery of "high-temperature" superconductivity in $(La_{1-x}Ba_x)_2CuO_4$ in the late 1986 [4] and the discovery and study of many more complex copper oxides ("high-T_C cuprates") that immediately followed.

The ongoing explosion of interest in 3d-transition metal oxides has produced further breakthroughs, especially the discovery of "colossal magnetoresistance" (CMR) in ternary manganites in the 1990's. These advances have been accompanied by an unprecedented shortening of the time lag between the initial discoveries of novel materials with surprising fundamental properties, and the development of derivative materials and hybrid structures for the marketplace. Furthermore, an intensified interplay between basic research and information technologies has led to the creation of whole new fields of investigation and commercial development, including "spintronics", "magnonics", "multiferroics", "nano-science", "quantum computing", etc. These remarkable developments have been accompanied by an explosion of technical publications, which has completely outstripped the steady increases in journal publications on condensed matter physics and physical chemistry which took place in the 1960s and 1970s.

It is now apparent that novel quantum materials that exhibit surprising or even revolutionary physical properties are often the basis for technical breakthroughs that decisively influence the everyday lives of average people. Materials scientists, confronted with ever-increasing pace and competition in research, are beginning to examine the remaining "unknown territories" located in the lower rows of the periodic table of the elements. Although the rare earth and light actinide elements have been aggressively studied for many decades, the 4d- and 5d-elements and their oxides were largely ignored until recently (see **Table 1.1**, which gives common oxidation states and their corresponding ionic radii). The reduced abundance and increased production costs for many of these elements have certainly discouraged basic and applied research into their properties. Indeed, eight of the nine least-abundant stable elements in the Earth's crust are 4d- and 5d-transition metal elements (i.e., Ru, Rh, Pd, Re, Os, Ir, Pt, and Au; tellurium is the ninth).

Only since the late 1990s and the early 2000s has it become apparent that 4d- and 5d-transition metal oxides host unique competitions between fundamental interactions; these circumstances have yielded peculiar quantum states and empirical trends that markedly differ from conventional wisdom in general and those of their 3d counterparts in particular. It is not surprising that the 4d- and 5d-transition metal oxides exhibit not only every cooperative phase known in solids, but also feature novel quantum states seldom or never seen in other materials.

1.2 Outline of the Book

This book is intended for graduate students and other materials scientists who have a background in condensed matter or materials physics. There are distinct structural and chemical characteristics of $4d$- and $5d$-transition metal oxides; these are lucidly and thoroughly discussed in the monograph *Transition Metal Oxides* by P.A. Cox (1995) [5]. The book describes a wide range of phenomena displayed by transition metal oxides, particularly the $3d$-transition metal oxides. The book also discusses various theoretical models proposed to interpret their physical properties. It is an accessible, resourceful reference for materials research. The present book focuses on the distinct, or even unique, physical properties of the $4d$- and $5d$-transition metal oxides that have been discovered and/or studied since the publication of Cox's monograph. The reader is cautioned that this book is by no means an exhaustive account of such a diverse, rapidly developing field, and reflects the interests of the authors.

This book is divided into three parts: Fundamental Principles, Novel Phenomena in $4d$- and $5d$-Transition Metal Oxides, and Single-Crystal Synthesis.

1.2.1 Chapter 1: Fundamental Principles

Chapter 1 focuses on underlying electronic interactions and structural or orbital effects that are important in determining the physical properties of $4d$- and $5d$-transition metal oxides. Many of these aspects are thoroughly discussed in *Transition Metal Oxides* [5], *Transition Metal Compounds* by Daniel Khomskii [6], and a review article on metal-insulator transitions [7]. Therefore, Chapter 1 emphasizes topics identified as most relevant to particularly unusual behaviors of $4d$- and $5d$-transition metal oxides.

1.2.2 Chapters 2–5: Novel Phenomena in $4d$- and $5d$-Transition Metal Oxides

Chapters 2–5 form the central part of this book, in which various physical behaviors unique to these materials are presented and discussed in detail. In particular, the SOI-driven Mott state constitutes a new quantum state that occurs within a class of correlated insulators that occupy the strong SOI limit [8–10]. Various topological and quantum spin-liquid states are proposed in view of the strong SOI extant in these materials, especially in $5d$-electron iridates [11–21]. Unlike in the $3d$-transition metal oxides, superconductivity is not a common occurrence in the $4d$- and $5d$-transition metal oxides, but when it occurs, it is extraordinary. For example, the pair symmetry of superconducting Sr_2RuO_4 [22] was initially thought to have a p-wave symmetry [23], but this characterization has been challenged in recent years [24–26], and the true nature of this superconductor is still open to debate after more than a quarter century has passed since its discovery in 1994.

The extended nature of $4d$- and $5d$-orbitals and SOI lead to physical properties that tend to be strong functions of lattice degrees of freedom. Virtually any external stimuli that readily couple to the lattice can easily induce novel phenomena, such as electrical-current-controlled states in Ca_2RuO_4 and Sr_2IrO_4 [27–30].

The critical, unique role the lattice plays in $4d$- and $5d$-transition metal oxides is also illustrated in the response to application of pressure. It is commonly anticipated that an insulating state will collapse in favor of an emergent metallic state at high pressures since the average electron density must increase with pressure, while the electronic bandwidth is expected to broaden and fill the insulating energy band gap. This is true for $3d$-transition metal oxides and other materials where SOI is negligible or relatively weak. However, most iridates avoid metallization under high pressure; the most notable example is a persistent insulating state up to 185 GPa in Sr_2IrO_4 [31]. Pressure-induced structural distortions prevent the expected onset of metallization, despite the sizable volume compression attained at the highest pressure. More often than not, the ground states of these materials are on the verge of a phase transition or show borderline behavior. Consequently, contradictory physical phenomena can occur in the same material. For example, $Ca_3Ru_2O_7$ exhibits conflicting hallmarks of both insulating and metallic states that include unusual colossal magnetoresistivity and quantum oscillations periodic in both $1/B$ and B (B is magnetic induction). More recently, a quantum liquid in an unfrustrated square lattice $Ba_4Ir_3O_{10}$ readily transforms into a robust antiferromagnet with a slight lattice modification [32,33]. Magnetotransport properties of the $4d$- and $5d$-transition metal oxides tend to be dictated by the lattice and/or orbital degrees of freedom because the spins are coupled to the lattice via strong SOI [34,35]. This contrasts with the $3d$-transition metal oxides, which are dominated by the spin degree of freedom alone.

1.2.3 Chapter 6: Single-Crystal Synthesis

One challenge that experimentalists confront in studying $4d$- and $5d$-transition metal oxides is materials synthesis, especially single-crystal growth, because of the high vapor pressure and high melting points exhibited by these materials. The challenge becomes even more daunting because most of these materials are inherently distorted, to which their physical properties are extremely susceptible. Chapter 6 focuses on a few single-crystal techniques that are most suitable for the $4d$- and $5d$-transition metal oxides. This chapter also introduces a promising "field-altering technology" applied to improve sample quality during crystal growth at high temperatures [33]. This technology knowingly takes advantage of strong SOI to address a major challenge to today's research community: a great deal of theoretical work predicts novel quantum states that have thus far met very limited experimental confirmation. We believe this is chiefly due to the extreme susceptibility of spin-orbit-coupled materials to structural distortions/disorder [36]. This chapter is also intended to help fill an existing void in the literature describing relevant synthesis techniques for $4d$- and $5d$-materials.

1.3 Fundamental Characteristics of 4d- and 5d-Electron Transition Metal Oxides

One key characteristic of the $4d$- and $5d$-electron transition elements is that their d-orbitals are more extended compared to those of their $3d$-electron counterparts, as shown in **Fig. 1.1** [37]. Consequently, strong p-d hybridization and electron-lattice coupling,

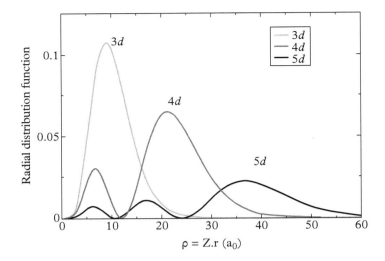

Fig. 1.1 *Radial distribution function as a function of spatial extent for 3d-, 4d-, and 5d-orbitals; Z is the effective atomic number and r is the distance from the nucleus [37].*

Table 1.2 *Comparison between 3d- and 4d/5d-Electrons*

Electron	U (eV)	λ_{so} (eV)	J_H (eV)	Key Interactions	Phenomena
3d	5-7	0.01-0.1	0.7-0.9	$U \gg J_H > \lambda_{so}$	HTSC/CMR
4d	0.5-3	0.1-0.3	0.5-0.6	$U > J_H > \lambda_{so}$	Orbital Order
5d	0.4-2	0.1-1	~0.5	$U \sim J_H \sim \lambda_{so}$	$J_{eff} = 1/2$ State

HTSC = High-temperature superconductivity; CMR = Colossal magnetoresistivity.

along with a reduced (with respect to 3d-transition metals) intra-atomic Coulomb interaction U, the Hund's rule coupling J_H and enhanced crystalline electric fields, are expected in these systems. Furthermore, materials containing these elements with high-atomic number Z exhibit scalar relativistic effects [38, 39] and strong SOI, which affect the total energy and thermodynamic stability of 4d- and 5d-materials. The SOI is known to exert only negligible effects on total energy and stability in 3d-materials (see **Table 1.2**). The strength of the SOI is expected to increase significantly in 4d- and 5d-transition metal compounds, since the SOI scales with Z^2 [40] (not the Z^4-dependence that is often cited in the literature). This is because the screening of the nuclear charge by the core electrons yields the effective Z^2-dependence of the SOI for the outer-shell electrons; the Z^4-dependence of SOI is appropriate only for unscreened hydrogenic wavefunctions. In any case, the phenomenology of the SOI and its fundamental consequences for material properties have been neglected until recently, due to the pervasive emphasis that has been placed upon the $3d$-elements. It is therefore appropriate to emphasize an unusual interplay between the competing interactions present in the $4d$- and $5d$-oxides, as it offers wide-ranging opportunities for the discovery of new physics and, ultimately, new device paradigms. The unique opportunities offered by the $4d$- and $5d$-transition elements are exemplified by novel phenomena only recently observed in $4d$-based ruthenates and $5d$-based iridates, which are the central focus of this book.

1.4 Crystal Fields and Chemical Aspects

The crystalline electric field is an electric field originating from neighboring ions in a crystal lattice; therefore the strength and symmetry of the crystal field sensitively depends on the symmetry of the local environment. Octahedral and tetrahedral environments are two common cases in which the former arrangement is more common than the latter. This is because most oxides have either perovskite or pyrochlore structures in which octahedra are the building blocks. Note that perovskites include various structural types, including hexagonal perovskites, e.g., $BaRuO_3$ and $BaIrO_3$ [41]. The crystal field generated by the near-neighbors at the M-site in MO_6 octahedra lifts the energy degeneracy of a free M-ion, splitting d-orbitals into two groups, namely, t_{2g} orbitals (d_{xy}, d_{yz}, and d_{xz}) and e_g orbitals (d_{z^2} and $d_{x^2-y^2}$) [42]. The e_g orbitals directly point toward the p-orbitals of the oxygen anions and experience a strong repulsion, thus shifting them to a higher energy with respect to that of the free ion. In contrast, the t_{2g} orbitals point between the p-orbitals of the oxygen anions and undergo a weaker repulsion, shifting them to a lower energy with respect to that of the free ion, as illustrated in **Fig. 1.2**. In a tetrahedral environment, the t_{2g} orbitals shift to higher energy than the e_g orbitals due to differences in orbital overlap between d- and p-electrons. Naturally, any structural distortions of either MO_6 octahedra or MO_4 tetrahedra further lifts the degeneracy of the t_{2g} and e_g orbitals. Perturbation theory of the Stark effect dictates the crystal field splitting (or energy shift), Δ, of the 6-fold t_{2g} orbitals and 4-fold e_g orbitals that will be - 4Dq and + 6Dq, respectively, totaling 10 Dq (**Fig. 1.2**). The parameter D is proportional to the nuclear electric charge Ze, where Z is the atomic number and e is the charge of an electron. The parameter q scales with r^4, in which r is the mean value of the radial distance between a d-electron and the nucleus. As a result, the strength of the crystal field increases with Z, and obeys the following order for $3d$-, $4d$-, and $5d$-electrons:

$$\Delta(3d) < \Delta(4d) \sim \Delta(5d).$$

Therefore, the orbital splitting between t_{2g} and e_g bands is greater in $4d$- and $5d$-oxides ($\Delta \sim$ 2–5 eV) than $3d$-oxides ($\Delta \sim$ 1–2 eV). Consequently, complete filling of the lowest-energy d-orbitals, rather than single occupation of the d-orbitals from the lowest energy and upward, is energetically favored. As such, the first Hund's rule, which requires an

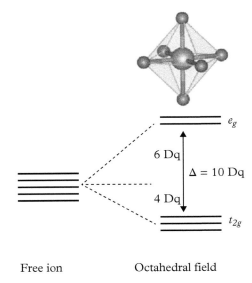

e_g

6 Dq

$\Delta = 10$ Dq

4 Dq

t_{2g}

Free ion   Octahedral field

Fig. 1.2 *The crystal field in an octahedral field. The crystal field splitting Δ arises when d-orbitals are placed in the octahedral field, and vanishes in a spherical symmetry.*

electronic wavefunction arrangement to maximize the spin S in order to minimize the on-site Coulomb interaction, often breaks down, leading to low-spin states. A low-spin state is also favored because of the reduced on-site Coulomb interaction U in 4d- and 5d-transition metal oxides compared to their 3d-counterparts.

A trend to higher oxidation states in 4d- and 5d-elements is more obvious than in 3d-elements. The ionic radius is a strong function of oxidation state, decreasing as the oxidation state increases or d-electrons are removed. In addition, the increase in the ionic radius from 3d- to 4d-elements is significant, but this increase is less noticeable when moving from 4d- to 5d-elements of the periodic table, primarily due to the screening provided by a full 4f-shell at the beginning of the 5d-row of the periodic table; for example, the ionic radii for tetravalent Co^{4+}, Rh^{4+}, and Ir^{4+} ions are 53, 60, and 62.5 pm, respectively (see **Table 1.1**). **Table 1.1** lists common oxidation states observed in transition metal oxides and their corresponding ionic radii.

1.5 Electron-Lattice Coupling

The extended nature of 4d- and 5d-electron orbitals generates strong electron-lattice couplings and crystalline electric field strengths in the "heavy" transition metal oxides. For example, the local environment of the Ru-ions determines the strong crystalline electric field splitting of the 4d-levels and, hence, the band structure of a given compound. As a result, structurally driven phenomena are commonplace in 4d- and

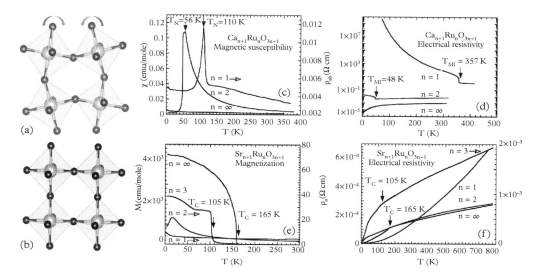

Fig. 1.3 *The comparison of the basal plane of RuO$_6$ octahedra between (a) Ca$_{n+1}$Ru$_n$O$_{3n+1}$ and (b) Sr$_{n+1}$Ru$_n$O$_{3n+1}$. Note that RuO$_6$ octahedra of Ca$_{n+1}$Ru$_n$O$_{3n+1}$ are severely rotated and tilted. Top panels: The temperature dependence of (c) the magnetic susceptibility χ and (d) the resistivity ρ for Ca$_{n+1}$Ru$_n$O$_{3n+1}$ with n = 1, 2, and ∞. Lower panels: The temperature dependence of (e) the magnetization M and (f) the resistivity ρ for Sr$_{n+1}$Ru$_n$O$_{3n+1}$ with n = 1, 2, 3, and ∞. Note the sharp differences in the ground state between the RP series of Ca- and Sr-compounds and n-dependence of χ, M, and ρ. T$_{MI}$ = metal-insulator transition; T$_N$ = Néel temperature; T$_C$ = Curie temperature.*

5d-oxides, and slight structural alterations may cause drastic changes in physical properties. A good example is the Ruddlesden-Popper (RP) series, Ca$_{n+1}$Ru$_n$O$_{3n+1}$ and Sr$_{n+1}$Ru$_n$O$_{3n+1}$, where n is the number of Ru-O layers per unit cell. The physical properties of this class of materials are very sensitive to the distortions and relative orientations of corner-shared RuO$_6$ octahedra. Effects of the ionic radius of the alkaline earth cation, which is 100 pm and 118 pm for Ca and Sr, respectively, must also be considered: the significantly smaller ionic radius of Ca causes severe structural distortions and rotations/tilts of RuO$_6$ octahedra in Ca$_{n+1}$Ru$_n$O$_{3n+1}$ (**Figs. 1.3a and 1.3b**), which produces ground states that are fundamentally different from those in Sr$_{n+1}$Ru$_n$O$_{3n+1}$. Due to the larger structural distortions for the Ca$_{n+1}$Ru$_n$O$_{3n+1}$ compounds, they are all proximate to a metal-insulator transition and prone to antiferromagnetic (AFM) order (see **Figs. 1.3c and 1.3d**), whereas the less-distorted Sr$_{n+1}$Ru$_n$O$_{3n+1}$ compounds are metallic and tend to be ferromagnetic (FM) (see **Figs. 1.3e and 1.3f**).

The observed trends for the magnetic ordering temperature with respect to the number of directly coupled Ru-O layers n is surprisingly different between these two isostructural and isoelectronic systems. The Curie temperature T$_C$ *increases* with n for Sr$_{n+1}$Ru$_n$O$_{3n+1}$, whereas the Néel temperature T$_N$ *decreases* with n for Ca$_{n+1}$Ru$_n$O$_{3n+1}$, as shown in **Fig. 1.4**. A semiquantitative ranking of W/U ratios can be created for these compounds according

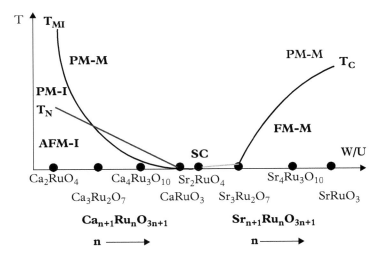

Fig. 1.4 *Phase diagram (T vs. W/U) qualitatively describing $Ca_{n+1}Ru_nO_{3n+1}$ and $Sr_{n+1}Ru_nO_{3n+1}$. Note the ground state can be readily changed by changing the cation, and physical properties can be systematically tuned by altering the number of Ru-O layers, n. SC = superconductor; FM-M = ferromagnetic metal; AFM-I = antiferromagnetic insulator; PM-M = paramagnetic metal [43].*

to their properties such that the two RP series can be placed into one phase diagram. Note that W stands for electronic bandwidth. Such a stark dependence of the ground state on the cation species (alternatively, the ionic radius) has not been observed in the $3d$-RP systems. Clearly, the lattice and orbital degrees of freedom play critical roles in the behavior of the $4d$- and $5d$-materials, which is corroborated by the wide array of intriguing phenomena and numerous novel phases (e.g., orbital ordering, orbitally driven colossal magnetoresistance, structurally driven Mott transition, lattice-driven magnetoresistance, etc.) that have been revealed under external stimuli coupling to the lattice [43].

1.6 Spin-Orbit Interactions

Besides the strong electron-lattice coupling, a strong SOI also drives the physical properties of $4d$- and $5d$-transition metal oxides. The SOI is a relativistic phenomenon arising from an interaction between the spin and orbital parts of an electron's wave function in an atom and is particularly strong in the $5d$-transition metal oxides, due to their relatively high $5d$-electron velocities. If we consider a comoving inertial frame in which the nucleus orbits an electron at rest, the orbiting nucleus generates a current, and thus a magnetic field that interacts with the spin of the electron. This interaction gives rise to a term in Hamiltonian proportional to $\mathbf{S \cdot L}$, where $\mathbf{S}$ is the spin angular momentum and $\mathbf{L}$ is the orbital angular momentum. The SOI scales with Z^4 in a hydrogenic atom but with Z^2 in a solid because of the screening of the nuclear charge by the core electrons, as discussed earlier [38–40]. Indeed, the SOI of *d*- and *f*-electrons scales much better with Z^2 than with Z^4, as illustrated in **Fig. 1.5**.

One important consequence of the SOI is that the wave functions of 5*d*-electrons become a coherent superposition of different orbital and spin states, leading to a peculiar distribution of spin densities in real space, as illustrated in **Fig. 1.6** [12]. A key result of the SOI is that the exchange Hamiltonian sensitively depends on structural details such as bond angles and bond lengths, one of central points of this book.

Traditional arguments suggest that 5*d*-transition metal oxides should be more metallic and less magnetic than materials based upon 3*d*-, 4*f*-, or even 4*d*-elements, because 5*d*-electron orbitals are more extended in space, which leads to increased electronic bandwidth. It is traditionally anticipated that the orbital interaction that generates electronic bandwidth W should follow the order of $W(5d) > W(4d) > W(3d) > W(4f)$ [5]. Studies in the later 1990s found that this expectation conflicts with two trends observed in RP iridates such as $Sr_{n+1}Ir_nO_{3n+1}$ (n = 1 and 2) and the hexagonal perovskite $BaIrO_3$ [36]. First, complex magnetic states occur with relatively high critical

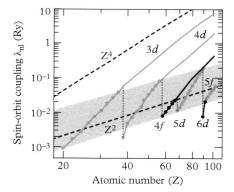

Fig. 1.5 *The dependence of the SOI on the atomic number Z. The outmost electrons of 3d, 4d, 5d, 4f and 5f are marked in the shaded area. The SOI scales better with Z^2 than Z^4. The upper dashed line is calculated results of Herman and Skillman for comparison for 3d electrons [39]; the lower dashed line is based on the Landau and Lifshitz scaling of SOI ~ Z^2 [40].*

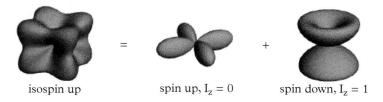

isospin up   spin up, $I_z = 0$ spin down, $I_z = 1$

Fig. 1.6 *An exemplary spin-density profile resulting from a hole in the isospin up state. It is a superposition of a different orbital and spin states [12].*

Table 1.3 *Exemplary Iridates*

System	Néel Temperature (K)	Ground State
Sr_2IrO_4 (n = 1)	240	Canted AFM insulator
$Sr_3Ir_2O_7$ (n = 2)	285	AFM insulator
$BaIrO_3$	183	Canted AFM insulator

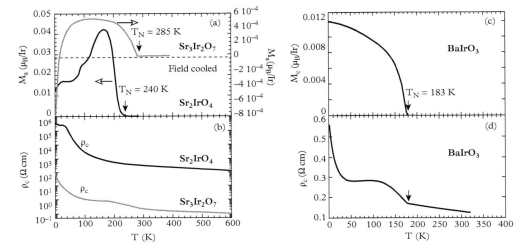

Fig. 1.7 *Iridates: Temperature dependence of (a) the a-axis magnetization M_a, (b) the c-axis resistivity ρ_c for Sr_2IrO_4 (black) and $Sr_3Ir_2O_7$ (gray), (c) the c-axis magnetization M_c and (d) the c-axis resistivity ρ_c for $BaIrO_3$.*

temperatures (up to 285 K), but with unusually low ordered moments. Second, "exotic insulating states" are observed rather than metallic states, as shown in **Fig. 1.7** and **Table 1.3** [44–48]. These anomalies did not draw much attention until 2008, when it was recognized that the critical underpinning of these unanticipated states is a strong SOI that vigorously competes with Coulomb interactions, crystalline electric fields, and Hund's rule coupling.

In general, the energy splitting depends on the ratio of the crystal field Δ to λ_{so} or Δ/λ_{so}. For a free ion or a spherical symmetry with $\Delta = 0$, the SOI splits the degenerate d-orbitals into a low-energy J = 3/2 quartet and a high-energy J = 5/2 multiplet. With a finite tetragonal crystal field Δ, a combined effect of Δ and λ_{so} further rearranges the orbitals, leading to $J_{eff} = 1/2$ and $J_{eff} = 3/2$ bands when $\Delta/\lambda_{so} > 1$ [8,37].

The so-called $J_{eff} = 1/2$ Mott state is a novel quantum state that served as an early example of the unique consequences of the strong SOI in iridates [8–10]. The SOI

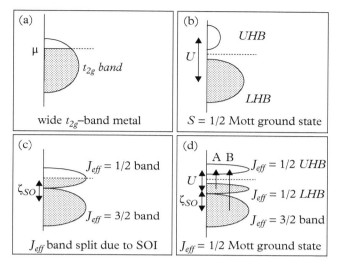

Fig. 1.8 *Traditional Mott insulator (b) originating from t_{2g} band (a). New type of $J_{eff} = 1/2$ insulator (d) originating from a split-off $J_{eff} = 1/2$ state (c) due to a combined effect of strong SOI and U [8].*

has an approximate strength of 0.4 eV in $5d$-iridates (compared to around 20 meV in $3d$-materials), which is strong enough to split the t_{2g} $5d$-bands into states with $J_{eff} = 1/2$ and $J_{eff} = 3/2$, the latter having lower energy (see **Table 1.2**). Since Ir^{4+} ($5d^5$) ions usually accommodate five $5d$-electrons in bonding states, we expect four to completely fill the lower $J_{eff} = 3/2$ bands, and one to partially fill the $J_{eff} = 1/2$ band in which the Fermi level E_F resides. The $J_{eff} = 1/2$ band is so narrow that even a reduced on-site Coulomb repulsion (U ~ 0.5 eV, due to the extended nature of $5d$-electron orbitals) is sufficient to open a small energy gap Δ_c that stabilizes an insulating state, as shown in **Fig. 1.8**.

The splitting between the $J_{eff} = 1/2$ and $J_{eff} = 3/2$ bands narrows as the dimensionality (i.e., n) increases in $Sr_{n+1}Ir_nO_{3n+1}$, and the two bands progressively broaden and build up a finite density of states near the Fermi surface, a trend that also characterizes the RP ruthenates discussed earlier. In particular, the bandwidth W of the $J_{eff} = 1/2$ band increases from 0.48 eV for n = 1, to 0.56 eV for n = 2, and 1.01 eV for n → ∞ (see **Fig. 1.9**) [9]. The ground state evolves, with decreasing Δ_c, from a robust insulating state for Sr_2IrO_4 (n = 1) to a metallic state for $SrIrO_3$ (n → ∞). A well-defined, but "marginal" insulating state for $Sr_3Ir_2O_7$ lies between them at n = 2 and at the border between a collinear AFM insulator and a spin-orbit-coupled Mott insulator.

The SOI can be modified by correlations among band electrons, that is, interatomic Coulomb interactions can be screened by itinerant band electrons. It has been suggested that Coulomb correlations can actually enhance the SOI in $4d$-electron systems such as Sr_2RhO_4 [49].

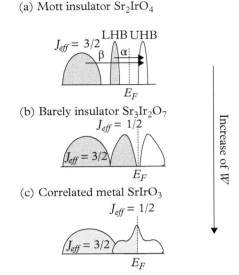

Fig. 1.9 Schematic *band diagrams of* $Sr_{n+1}Ir_nO_{3n+1}$: *(a)* Sr_2IrO_4 *(n = 1), (b)* $Sr_3Ir_2O_7$ *(n = 2), and (c)* $SrIrO_3$ *(n → ∞)*. E_F = *Fermi level; the arrow indicates a trend of the bandwidth W widening with increasing n* [9].

It is worth mentioning that Sr_2RhO_4 is similar to Sr_2RuO_4 and Sr_2IrO_4, both electronically and structurally, but its ground state is fundamentally different from those of the other two compounds. Sr_2RhO_4 hosts an Rh^{4+} ion with five $4d$-electrons (compared to four $4d$-electrons of the Ru^{4+} ion in Sr_2RuO_4). It shares a crystal structure remarkably similar to that of Sr_2IrO_4; in particular, the RhO_6 octahedron rotates about the c axis by 10°; this value is zero for Sr_2RuO_4 and 12° for Sr_2IrO_4. It is argued that this octahedral rotation facilitates a correlation-induced enhancement of the SOI by about 20%, that is, the SOI increases from a "bare value" of 0.16 eV to 0.19 eV [49]. Despite its similarities to insulating Sr_2IrO_4, Sr_2RhO_4 is a paramagnetic metal because the SOI is still not strong enough to conspire with the Coulomb interaction to open an energy gap [50]. On the other hand, Sr_2RhO_4 is indeed near to an insulating state because of the octahedral rotation and the enhanced SOI: slight Ir doping for Rh induces an insulating state [51]. The metallic state is less robust because the t_{2g} bands near the Fermi surface are less dispersive in Sr_2RhO_4 than in Sr_2RuO_4, and therefore more susceptible to SOI-induced band shifts near the Fermi surface than in Sr_2RuO_4 [52].

A great deal of theoretical and experimental work has appeared in response to early predictions of a large number of novel effects in spin-orbit-coupled systems: superconductivity, Weyl semimetals with Fermi arcs, axion insulators with strong magnetoelectric coupling, topological insulators, correlated topological insulators with large gaps enhanced by Mott physics, Kitaev modes, 3-D spin liquids with Fermionic spinons, topological semimetals, and Kitaev spin liquids [11–21]. However, most of these novel

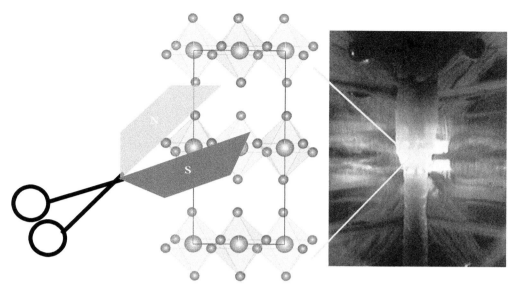

Fig. 1.10 *Field-altering technology: A schematic for field-altering a crystal structure (left) during crystal growth in the molten zone (right).*

states exist only in theoretical models and have thus far met very limited experimental confirmation. It is now recognized that the absence of the theoretically predicted states is due in part to the extreme susceptibility of relevant $4d$- and $5d$-materials to structural distortions and disorder [36,53,54]. The ground states of these materials are dictated by a delicate interplay between spin-orbit and Coulomb interactions, and slight perturbations, such as distortions/disorder, can provoke what appear to be disproportionate responses in physical properties, which is in sharp contrast to the situation in $3d$-transition metal oxides.

Considerations of the vulnerability of $4d$- and $5d$-materials to structural nuances must be extended to the synthetic processes needed to fabricate single-crystal sample materials. To fundamentally address this challenge, a "field-altering technology" is proposed to "correct" distortions and disorder by applying a strong magnetic field during crystal growth. A schematic of this technology is illustrated in **Fig. 1.10**. This technology is found to be extremely effective for spin-orbit-coupled materials [33] and is discussed in detail in Chapter 6.

1.7 The Dzyaloshinsky-Moriya Interaction

Spin canting is a common occurrence, and therefore an important consideration in understanding the physics of $4d$- and $5d$-transition metal oxides. The Dzyaloshinsky-Moriya (DM) interaction, or antisymmetric exchange, provides a simple model

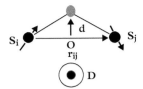

Fig. 1.11 *Schematic of the Dzyaloshinsky-Moriya interaction between two neighboring canted spins* S_i *and* S_j *at sites separated by a vector* r_{ij}. *The vector* **d** *is a measure of the displacement of the oxygen ion from the center* **O** *between the two magnetic ions. The vector* **D** *points out the page.*

Hamiltonian describing spin canting. It arises from the SOI, which plays an intermediate role similar to that of oxygen that facilitates superexchange interaction [55]. The DM interaction commonly operates in canted antiferromagnets such as α-Fe_2O_3, $Ca_3Ru_2O_7$, and Sr_2IrO_4. The standard DM Hamiltonian describes neighboring canted spins S_i and S_j, according to the relation,

$$H = D \cdot (S_i \times S_j), \text{ where } D = r_{ij} \times d \tag{1.1}$$

where S_i and S_j are the spins of two neighboring ions at i and j sites separated by the vector r_{ij}, as shown in **Fig. 1.11**. The vector **D**, which is directly related to the vector **d**, vanishes when **d** becomes zero (i.e., the center of inversion is recovered at the center point **O**) (**Fig. 1.11**). As Eq. 1.1 implies, the anisotropic exchange interaction favors a spin canting configuration so that the resultant vector of $S_i \times S_j$ (pointing into the page) is antiparallel to the vector **D** (pointing out of the page), giving rise to a negative energy—the greater the canting, the more negative the DM interaction becomes (see **Fig. 1.11**). The DM interaction therefore models weak ferromagnetic behavior in a canted antiferromagnet. Because a strong SOI tightly couples magnetic moments to the lattice, weak ferromagnetism is a common occurrence in $4d$- and $5d$-transition metal oxides having distorted structures. In fact, canted antiferromagnetism is particularly common among iridates [36].

1.8 Phase Diagram for Correlated, Spin-Orbit-Coupled Matter

The on-site Coulomb interaction U plays a dominant role in the physics of $3d$-transition metal oxides. However, there is no apparent dominant interaction in $4d$- and $5d$-transition metal oxides, especially in $5d$-transition metal oxides, where all relevant interactions are of comparable energy scales, and therefore strongly compete or cooperate (see **Table 1.2**). In particular, the SOI λ_{so} strongly competes with the electron-lattice and on-site Coulomb interaction U; therefore, any tool that allows one to tune the relative strengths of λ_{so} and U has a potential to offer an opportunity to discover and study novel

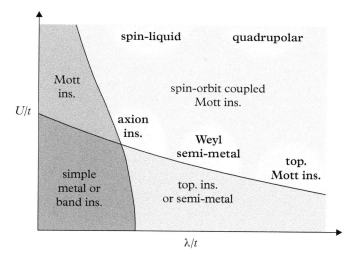

Fig. 1.12 *A schematic phase diagram in terms of Coulomb interaction U and SOI* λ_{so} *[11]*.

materials ranging from Mott insulators ($U > \lambda_{so}$), topological semimetals ($\lambda_{so} > U$), and SOI-driven Mott insulators ($\lambda_{so} \sim U$), to strong SOI-driven topological states ($\lambda_{so} \geq U$), as schematically illustrated in **Fig. 1.12**. Studies suggest that the Coulomb and spin-orbit interactions can cooperate to "bootstrap" stable insulating states [11]: on one hand, narrow bands generated by an increased SOI become more susceptible to localization via U; on the other hand, electron localization is enhanced with increasing U, which in turn enhances the relative strength of the SOI.

Studies of $4d$- and $5d$-transition metal oxides have already uncovered a large array of new phenomena that depend on the existence of unusual interplays between available electronic interactions. Several extensively studied systems offer a glimpse of novel physics: unconventional superconductivity below $T_C = 1.5$ K in Sr_2RuO_4 [22,26], colossal negative thermal expansion via magnetic and orbital orders in doped Ca_2RuO_4 [56–58], contradictory physical properties and extreme anisotropy in $Ca_3Ru_2O_7$ [43,59,60], field-induced quantum criticality in $Sr_3Ru_2O_7$ [61], orbital order in $La_4Ru_2O_{10}$ [62], a possible ferroelectric/metallic state in $LiOsO_3$ [63] and $Cd_2Re_2O_7$ [64–65] (it is also known for its metal-insulator transition at 226 K and superconductivity below 2 K [66,67]), SOI-driven $J_{eff} = 1/2$ state in Sr_2IrO_4, possible quantum spin-liquid states in $Na_4Ir_3O_8$, Na_2IrO_3, and Li_2IrO_3 [20,68,70–79]. More recent observations include exotic magnetism in double-perovskite Sr_2YIrO_6 [80], and a quantum liquid in an unfrustrated square lattice, $Ba_4Ir_3O_{10}$ [32].

Transition metal oxides arguably constitute the most diverse and fascinating group of quantum materials presently known [5,6,7,43,81,82]. Some representative $4d$- and $5d$-transition metal oxides and their major characteristic phenomena are compiled in **Table 1.4** (note that most oxidation states are tetravalent unless specifically marked). This list is by no means complete, as $4d$- and $5d$-transition metal oxides are the subject

Table 1.4 *Representative 4d- and 5d-Transition Metal Oxides and Physical Phenomena*

Compound	Major Phenomena
Ca_2RuO_4 and derivatives	Structurally driven Mott insulator with T_{MI} = 357 K and T_N = 110 K [56,83]; colossal negative thermal expansion via orbital and magnetic orders [57]; current-driven orbital state [29]
Sr_2RuO_4	Novel superconductor with T_C = 1.5 K [22]
$Ca_3Ru_2O_7$	Orbitally driven colossal magnetoresistivity; bulk spin valve; quantum oscillations periodic in 1/B and B; antiferromagnetic metallic state with T_N = 56 K, Mott transition with T_{MI} = 48 K [59,84,85]
$Sr_3Ru_2O_7$	Field-induced quantum criticality [61]
$Sr_4Ru_3O_{10}$	Strong magnetic anisotropy, coexistence of band-ferromagnetism (T_C = 105 K) and metamagnetism [86]
$CaRuO_3$	Enhanced paramagnetism [87]
$SrRuO_3$	Itinerant ferromagnetism with T_C = 165 K [87]
$La_4Ru_2O_{10}$	Orbital order [62]
$Na_2Ru_4O_9$	Anomalous quasi-one-dimensional behavior [88]
Li_2RuO_3	Antiferromagnetic insulator [89]
Na_2RuO_3	Antiferromagnetic insulator [89]
Sr_2YRuO_6 (Ru^{5+}:$4d^3$)	Antiferromagnetic insulator [90]
Sr_2RhO_4	Paramagnetic metal [91]
$Sr_5Rh_4O_{12}$	Partial antiferromagnetic order [92]
Li_2RhO_3	Spin-glassy relativistic Mott state [93]
Sr_2IrO_4	J_{eff} = 1/2 Mott state [8–10]; current-driven quantum state [27]
$Sr_3Ir_2O_7$	Borderline J_{eff} = 1/2 Mott state [47]
Sr_2YIrO_6, Ba_2YIrO_6 (Ir^{5+}:$5d^4$)	Exotic magnetic state driven by competing non-cubic crystal field and SOI; quantum fluctuations [80,94]

Table 1.4 *continued*

Compound	Major Phenomena
$Ca_5Ir_3O_{12}$	Partial antiferromagnetism [92]
Ca_4IrO_6	Partial antiferromagnetism [92]
$BaIrO_3$	Coexisting weak ferromagnetism and density waves [48]
$Ba_4Ir_3O_{10}$	Quantum liquid [32]
$Ba_{13}Ir_6O_{30}$ (Ir^{5+}, Ir^{6+})	Novel magnetism [95]
$Ba_5AlIr_2O_{11}$ (Ir^{4+}, Ir^{5+})	Coexisting charge and magnetic order [96]
Li_2IrO_3	Antiferromagnetism with $T_N = 15$ K [78]
Na_2IrO_3	Zigzag magnetic order with $T_N = 18$ K [74–77]
$Na_4Ir_3O_8$	Possible spin liquid state [97]
$NaIrO_3$ (Ir^{5+})	Correlated insulator [69]
$SrIrO_3$ (Hex)	Non-Fermi-liquid [98]
$SrIrO_3$ (Tetragonal)	Correlated metal [9]
$R_2Ir_2O_7$ (R = rare earth ion)	Metal-insulator transition; geometric frustration; quantum fluctuations [99]
$Bi_2Ir_2O_7$	Correlated metal with low-temperature magnetic order [100–101]
$Cd_2Re_2O_7$	Structurally driven transition at 226 K and superconductivity below 2 K [66–67]
$Cd_2Os_2O_7$	Nonlinear magnetism [102]
$R_2Mo_2O_7$	Metallic ferromagnetism, semiconducting spin glass [103]
$LiOsO_3$	Ferroelectric metal [63]
Ba_2NaOsO_6	Ferromagnetism in Mott state [104]
$NaOsO_3$	Continuous metal-insulator transition, antiferromagnetism [105]
KOs_2O_6	Superconductivity with $T_C = 9.6$ K [106]

of a rapidly growing field of research, but it already presents an impressive scope of novel physical phenomena.

1.9 Absence of Topological States in 4*d*- and 5*d*-Transition Metal Oxides

Spin-orbit-driven topological states and materials have been front and center in condensed matter research in recent years [107]. Almost all the currently known topological states arising from strong SOI-induced band inversion are found in selenides and tellurides [108]. There is so far no clear-cut materials realization of any topological states in spin-orbit-coupled transition metal oxides, despite extensive theoretical and experimental efforts. The absence of topological states in the oxides is fascinating. Certain differences in empirical trends between transition metal oxides and chalcogenides are obvious (see **Fig. 1.13**). The oxides are more electronically correlated, thus more insulating and magnetic, in part because of the four elements in the *VIA* column in the periodic table: oxygen has the strongest electronegativity with a tendency to attract electrons, thus the oxides tend to be more ionic. Since the electronegativity decreases following the order of O, S, Se, and Te, the ionic bonding is no longer as significant in heavier chalcogenides; instead, strong competition between metal-ligand and ligand-ligand bonding becomes a driver of structural complexity in chalcogenides. The underlying physical behavior systematically evolves from more insulating, magnetic states (Subgroup I. Oxides), to more metallic, nonmagnetic, often superconducting states (with $T_C < 5$ K) (Subgroup III. Tellurides), through a crossover regime where sulfides and selenides are situated (Subgroup II. Sulfides and Selenides) (**Fig. 1.13**). Orbital diamagnetism, whose strength is proportional to the atomic number Z and the ionic radius r, is often strong in tellurides under application of magnetic field. In addition, the energy gaps for topological insulators in chalcogenides tend to be narrow, and often less than 0.3 eV.

It is already established that physical phenomena in 4*d*- and 5*d*-materials are dictated by a combined effect of spin-orbit and Coulomb interactions. The stark differences in the empirical trends between oxides, sulfides, selenides, and tellurides stress the

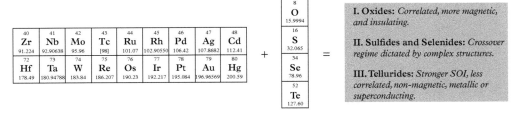

Fig. 1.13 *Transition metal oxides and chalcogenides: Quantum phenomena in 4d- and 5d-materials are dictated by combined effects of spin-orbit and Coulomb interactions. The relative strength of* λ_{so} *and U varies in the three sub-groups: (1) Oxides, (2) Sulfides and Selenides, (3) Tellurides.*

significance of the variable relative strength of λ_{so} and U in these materials. Apparently, electronic correlations are much weaker than the SOI (λ_{so} > U or λ_{so} >> U) in tellurides and, to a certain extent, in selenides, which may explain the fact that these materials are fertile ground for various topological states. In contrast, the electronic correlations are stronger (consequently, U > λ_{so} or U ~ λ_{so}) in oxides, which host all ordered states known in condensed matter. This trend is qualitatively consistent with predications shown in **Fig. 1.12**.

It is both challenging and interesting to systematically explore transition metal oxides and chalcogenides with varied relative strengths of λ_{so} and U. An obvious materials hurdle that needs to be crossed is that any new compound cannot be readily realized simply by substituting one element with the other among O, S, Se, and Te, because crystal structures of oxides tend to be different from those of chalcogenides, although sulfides and selenides are more likely to share the same structures, thus making them more chemically and structurally compatible.

Further Reading

- *Metal-Insulator Transitions*, Nevill Mott (Taylor & Francis, 1990)
- *Transition Metal Oxides*, P.A. Cox (Oxford, 1995)
- *Transition Metal Compounds*, Daniel Khomskii (Cambridge, 2014)
- *Metal-Insulator Transitions*, Masatoshi Imada, Atsushi Fujimori, and Yoshinori Tokura, *Reviews of Modern Physics* 70, 1039 (1998)

References

1. Hever, A., Oses, C., Curtarolo, S., Levy, O., Natan, A. The Structure and Composition Statistics of 6A Binary and Ternary Crystalline Materials, *Inorg. Chem.* **57**, 653 (2018)
2. Goodenough, J. B. Theory of the Role of Covalence in the Perovskite-Type Manganites [La, M(II)]MnO$_3$, *Phys. Rev.* **100**, 564 (1955)
3. Goodenough, J. B. An Interpretation of the Magnetic Properties of the Perovskite-Type Mixed Crystals La$_{1-x}$Sr$_x$CoO$_{3-\lambda}$, *Journal of Physics and Chemistry of Solids* **6**, 287 (1958)
4. Bednorz, J. G., Müller, K. A. Possible high T$_c$ superconductivity in the Ba−La−Cu−O system, *Z. Physik B - Condensed Matter* **64**, 189 (1986)
5. Cox, P. A. *Transition Metal Oxides: An Introduction to Their Electronic Structure and Properties.* (Oxford University Press, 1995)
6. Khomskii, D. *Transition Metal Compounds.* (Cambridge University Press, 2014)
7. Imada, M., Fujimori, A., Tokura, Y. Metal-insulator transitions, *Rev. Mod. Phys.* **70**, 1039 (1998)
8. Kim, B. J., Jin, H., Moon, S. J., Kim, J.-Y., Park, B.-G., Leem, C. S., Yu, J., Noh, T. W., Kim, C., Oh, S.-J., Park, J.-H., Durairaj, V., Cao, G., Rotenberg, E. Novel J$_{eff}$=1/2 Mott State Induced by Relativistic Spin-Orbit Coupling in Sr$_2$IrO$_4$, *Phys. Rev. Lett.* **101**, 076402 (2008)
9. Moon, S. J., Jin, H., Kim, K. W., Choi, W. S., Lee, Y. S., Yu, J., Cao, G., Sumi, A., Funakubo, H., Bernhard, C., Noh, T. W. Dimensionality-Controlled Insulator-Metal Transition and

Correlated Metallic State in $5d$ Transition Metal Oxides $Sr_{n+1}Ir_nO_{3n+1}$ (n=1, 2, and ∞), *Phys. Rev. Lett.* **101**, 226402 (2008)

10. Kim, B. J., Ohsumi, H., Komesu, T., Sakai, S., Morita, T., Takagi, H., Arima, T. Phase-Sensitive Observation of a Spin-Orbital Mott State in Sr_2IrO_4, *Science* **323**, 1329 (2009)

11. Witczak-Krempa, W., Chen, G., Kim, Y. B., Balents, L. Correlated Quantum Phenomena in the Strong Spin-Orbit Regime, *Annual Review of Condensed Matter Physics* **5**, 57 (2014)

12. Jackeli, G., Khaliullin, G. Mott Insulators in the Strong Spin-Orbit Coupling Limit: From Heisenberg to a Quantum Compass and Kitaev Models, *Phys. Rev. Lett.* **102**, 017205 (2009)

13. Pesin, D., Balents, L. Mott Physics and Band Topology in Materials with Strong Spin–Orbit Interaction, *Nature Phys* **6**, 376 (2010)

14. Wan, X., Turner, A. M., Vishwanath, A., Savrasov, S. Y. Topological Semimetal and Fermi-Arc Surface States in the Electronic Structure of Pyrochlore Iridates, *Phys. Rev. B* **83**, 205101 (2011)

15. Wang, F., Senthil, T. Twisted Hubbard Model for Sr_2IrO_4: Magnetism and Possible High Temperature Superconductivity, *Phys. Rev. Lett.* **106**, 136402 (2011)

16. Watanabe, H., Shirakawa, T., Yunoki, S. Monte Carlo Study of an Unconventional Superconducting Phase in Iridium Oxide J_{eff}=1/2 Mott Insulators Induced by Carrier Doping, *Phys. Rev. Lett.* **110**, 027002 (2013)

17. Meng, Z. Y., Kim, Y. B., Kee, H.-Y. Odd-Parity Triplet Superconducting Phase in Multiorbital Materials with a Strong Spin-Orbit Coupling: Application to Doped Sr_2IrO_4, *Phys. Rev. Lett.* **113**, 177003 (2014)

18. Yang, Y., Wang, W.-S., Liu, J.-G., Chen, H., Dai, J.-H., Wang, Q.-H. Superconductivity in Doped Sr_2IrO_4: A Functional Renormalization Group Study, *Phys. Rev. B* **89**, 094518 (2014)

19. You, Y.-Z., Kimchi, I., Vishwanath, A. Doping a Spin-Orbit Mott Insulator: Topological Superconductivity from the Kitaev-Heisenberg Model and Possible Application to $(Na_2/Li_2)IrO_3$, *Phys. Rev. B* **86**, 085145 (2012)

20. Chaloupka, J., Jackeli, G., Khaliullin, G. Kitaev-Heisenberg Model on a Honeycomb Lattice: Possible Exotic Phases in Iridium Oxides A_2IrO_3, *Phys. Rev. Lett.* **105**, 027204 (2010)

21. Rau, J. G., Lee, E. K.-H., Kee, H.-Y. Spin-Orbit Physics Giving Rise to Novel Phases in Correlated Systems: Iridates and Related Materials, *Annual Review of Condensed Matter Physics* **7**, 195 (2016)

22. Maeno, Y., Hashimoto, H., Yoshida, K., Nishizaki, S., Fujita, T., Bednorz, J. G., Lichtenberg, F. Superconductivity in a Layered Perovskite without Copper, *Nature* **372**, 532 (1994)

23. Mackenzie, A. P., Maeno, Y. The Superconductivity of Sr_2RuO_4 and the Physics of Spin-Triplet Pairing, *Rev. Mod. Phys.* **75**, 657 (2003)

24. Hicks, C. W., Brodsky, D. O., Yelland, E. A., Gibbs, A. S., Bruin, J. A. N., Barber, M. E., Edkins, S. D., Nishimura, K., Yonezawa, S., Maeno, Y., Mackenzie, A. P. Strong Increase of T_c of Sr_2RuO_4 Under Both Tensile and Compressive Strain, *Science* **344**, 283 (2014)

25. Steppke, A., Zhao, L., Barber, M. E., Scaffidi, T., Jerzembeck, F., Rosner, H., Gibbs, A. S., Maeno, Y., Simon, S. H., Mackenzie, A. P., Hicks, C. W. Strong Peak in T_c of Sr_2RuO_4 Under Uniaxial Pressure, *Science* **355**, 148 (2017)

26. Pustogow, A., Luo, Y., Chronister, A., Su, Y.-S., Sokolov, D. A., Jerzembeck, F., Mackenzie, A. P., Hicks, C. W., Kikugawa, N., Raghu, S., Bauer, E. D., Brown, S. E. Constraints on the Superconducting Order Parameter in Sr_2RuO_4 from Oxygen-17 Nuclear Magnetic Resonance, *Nature* **574**, 72 (2019)

27. Cao, G., Terzic, J., Zhao, H. D., Zheng, H., De Long, L. E., Riseborough, P. S. Electrical Control of Structural and Physical Properties via Strong Spin-Orbit Interactions in Sr_2IrO_4, *Phys. Rev. Lett.* **120**, 017201 (2018)

28. Bertinshaw, J., Gurung, N., Jorba, P., Liu, H., Schmid, M., Mantadakis, D. T., Daghofer, M., Krautloher, M., Jain, A., Ryu, G. H., Fabelo, O., Hansmann, P., Khaliullin, G., Pfleiderer, C., Keimer, B., Kim, B. J. Unique Crystal Structure of Ca_2RuO_4 in the Current Stabilized Semimetallic State, *Phys. Rev. Lett.* **123**, 137204 (2019)

29. Zhao, H., Hu, B., Ye, F., Hoffmann, C., Kimchi, I., Cao, G. Nonequilibrium Orbital Transitions via Applied Electrical Current in Calcium Ruthenates, *Phys. Rev. B* **100**, 241104(R) (2019)

30. Zhang, J., McLeod, A. S., Han, Q., Chen, X., Bechtel, H. A., Yao, Z., Gilbert Corder, S. N., Ciavatti, T., Tao, T. H., Aronson, M., Carr, G. L., Martin, M. C., Sow, C., Yonezawa, S., Nakamura, F., Terasaki, I., Basov, D. N., Millis, A. J., Maeno, Y., Liu, M. Nano-Resolved Current-Induced Insulator-Metal Transition in the Mott Insulator Ca_2RuO_4, *Phys. Rev. X* **9**, 011032 (2019)

31. Chen, C., Zhou, Y., Chen, X., Han, T., An, C., Zhou, Y., Yuan, Y., Zhang, B., Wang, S., Zhang, R., Zhang, L., Zhang, C., Yang, Z., DeLong, L. E., Cao, G. Persistent Insulating State at Megabar Pressures in Strongly Spin-Orbit Coupled Sr_2IrO_4, *Phys. Rev. B* **101**, 144102 (2020)

32. Cao, G., Zheng, H., Zhao, H., Ni, Y., Pocs, C. A., Zhang, Y., Ye, F., Hoffmann, C., Wang, X., Lee, M., Hermele, M., Kimchi, I. Quantum Liquid from Strange Frustration in the Trimer Magnet $Ba_4Ir_3O_{10}$, *npj Quantum Mater.* **5**, 26 (2020)

33. Cao, G., Zhao, H., Hu, B., Pellatz, N., Reznik, D., Schlottmann, P., Kimchi, I. Quest for Quantum States via Field-Altering Technology, *npj Quantum Mater.* **5**, 83 (2020)

34. Ye, F., Chi, S., Chakoumakos, B. C., Fernandez-Baca, J. A., Qi, T., Cao, G. Magnetic and Crystal Structures of Sr_2IrO_4: A Neutron Diffraction Study, *Phys. Rev. B* **87**, 140406(R) (2013)

35. Torchinsky, D. H., Chu, H., Zhao, L., Perkins, N. B., Sizyuk, Y., Qi, T., Cao, G., Hsieh, D. Structural Distortion-Induced Magnetoelastic Locking in Sr_2IrO_4 Revealed through Nonlinear Optical Harmonic Generation, *Phys. Rev. Lett.* **114**, 096404 (2015)

36. Cao, G., Schlottmann, P. The Challenge of Spin–Orbit-Tuned Ground States in Iridates: a Key Issues Review, *Rep. Prog. Phys.* **81**, 042502 (2018)

37. Martins, C., Aichhorn, M., Biermann, S. Coulomb Correlations in $4d$ and $5d$ Oxides from First Principles—or How Spin–Orbit Materials Choose Their Effective Orbital Degeneracies, *J. Phys.: Condens. Matter* **29**, 263001 (2017)

38. Landau, L. D., Lifshitz, E. M. *Quantum Mechanics: Non-Relativistic Theory*. (Pergamon Press, 1977)

39. Herman, F., Skillman, S. *Atomic Structure Calculations*. (Prentice-Hall, 1963)

40. Shanavas, K. V., Popović, Z. S., Satpathy, S. Theoretical Model for Rashba Spin-Orbit Interaction in d Electrons, *Phys. Rev. B* **90**, 165108 (2014)

41. Mitchell, R. H. *Perovskites: Modern and Ancient*. (Almaz Press, 2002)

42. Burns, R. G. *Mineralogical Applications of Crystal Field Theory*. (Cambridge University Press, 1993)

43. Cao, G., DeLong, L. *Frontiers of 4d- and 5d-Transition Metal Oxides*. (World Scientific, 2013)

44. Cava, R. J., Batlogg, B., Kiyono, K., Takagi, H., Krajewski, J. J., Peck, W. F., Rupp, L. W., Chen, C. H. Localized-to-Itinerant Electron Transition in $Sr_2Ir_{1-x}Ru_xO_4$, *Phys. Rev. B* **49**, 11890 (1994)

45. Crawford, M. K., Subramanian, M. A., Harlow, R. L., Fernandez-Baca, J. A., Wang, Z. R., Johnston, D. C. Structural and Magnetic Studies of Sr_2IrO_4, *Phys. Rev. B* **49**, 9198 (1994)

46. Cao, G., Bolivar, J., McCall, S., Crow, J. E., Guertin, R. P. Weak Ferromagnetism, Metal-to-Nonmetal Transition, and Negative Differential Resistivity in Single-Crystal Sr_2IrO_4, *Phys. Rev. B* **57**, R11039(R) (1998)

47. Cao, G., Xin, Y., Alexander, C. S., Crow, J. E., Schlottmann, P., Crawford, M. K., Harlow, R. L., Marshall, W. Anomalous Magnetic and Transport Behavior in the Magnetic Insulator $Sr_3Ir_2O_7$, *Phys. Rev. B* **66**, 214412 (2002)

48. Cao, G., Crow, J. E., Guertin, R. P., Henning, P. F., Homes, C. C., Strongin, M., Basov, D. N., Lochner, E. Charge Density Wave Formation Accompanying Ferromagnetic Ordering in Quasi-One-Dimensional $BaIrO_3$, *Solid State Communications* **113**, 657 (2000)

49. Liu, G.-Q., Antonov, V. N., Jepsen, O., Andersen., O. K. Coulomb-Enhanced Spin-Orbit Splitting: The Missing Piece in the Sr_2RhO_4 Puzzle, *Phys. Rev. Lett.* **101**, 026408 (2008)

50. Martins, C., Aichhorn, M., Vaugier, L., Biermann, S. Reduced Effective Spin-Orbital Degeneracy and Spin-Orbital Ordering in Paramagnetic Transition-Metal Oxides: Sr_2IrO_4 versus Sr_2RhO_4, *Phys. Rev. Lett.* **107**, 266404 (2011)

51. Qi, T. F., Korneta, O. B., Li, L., Butrouna, K., Cao, V. S., Wan, X., Schlottmann, P., Kaul, R. K., Cao, G. Spin-Orbit Tuned Metal-Insulator Transitions in Single-Crystal $Sr_2Ir_{1-x}Rh_xO_4$ ($0 \le x \le 1$), *Phys. Rev. B* **86**, 125105 (2012)

52. Haverkort, M. W., Elfimov, I. S., Tjeng, L. H., Sawatzky, G. A., Damascelli, A. Strong Spin-Orbit Coupling Effects on the Fermi Surface of Sr_2RuO_4 and Sr_2RhO_4, *Phys. Rev. Lett.* **101**, 026406 (2008)

53. Korneta, O. B., Qi, T., Chikara, S., Parkin, S., De Long, L. E., Schlottmann, P., Cao, G. Electron-Doped $Sr_2IrO_{4-\delta}$ ($0 \le \delta \le 0.04$): Evolution of a Disordered $J_{eff} = 1/2$ Mott insulator into an Exotic Metallic State, *Phys. Rev. B* **82**, 115117 (2010)

54. Ge, M., Qi, T. F., Korneta, O. B., De Long, D. E., Schlottmann, P., Crummett, W. P., Cao, G. Lattice-Driven Magnetoresistivity and Metal-Insulator Transition in Single-Layered Iridates, *Phys. Rev. B* **84**, 100402(R) (2011)

55. Moriya, T. Anisotropic Superexchange Interaction and Weak Ferromagnetism, *Phys. Rev.* **120**, 91 (1960)

56. Cao, G., McCall, S., Shepard, M., Crow, J. E., Guertin, R. P. Magnetic and Transport Properties of Single-Crystal Ca_2RuO_4: Relationship to Superconducting Sr_2RuO_4, *Phys. Rev. B* **56**, R2916(R) (1997)

57. Qi, T. F., Korneta, O. B., Parkin, S., De Long, L. E., Schlottmann, P., Cao, G. Negative Volume Thermal Expansion Via Orbital and Magnetic Orders in $Ca_2Ru_{1-x}Cr_xO_4$ ($0 < x < 0.13$), *Phys. Rev. Lett.* **105**, 177203 (2010)

58. Qi, T. F., Korneta, O. B., Parkin, S., Hu, J., Cao, G. Magnetic and Orbital Orders Coupled to Negative Thermal Expansion in Mott Insulators $Ca_2Ru_{1-x}M_xO_4$ (M = Mn and Fe), *Phys. Rev. B* **85**, 165143 (2012)

59. Cao, G., McCall, S., Crow, J. E., Guertin, R. P. Observation of a Metallic Antiferromagnetic Phase and Metal to Nonmetal Transition in $Ca_3Ru_2O_7$, *Phys. Rev. Lett.* **78**, 1751 (1997)

60. Cao, G., Lin, X. N., Balicas, L., Chikara, S., Crow, J. E., Schlottmann, P. Orbitally Driven Behaviour: Mott Transition, Quantum Oscillations and Colossal Magnetoresistance in Bilayered $Ca_3Ru_2O_7$, *New J. Phys.* **6**, 159 (2004)

61. Grigera, S. A., Perry, R. S., Schofield, A. J., Chiao, M., Julian, S. R., Lonzarich, G. G., Ikeda, S. I., Maeno, Y., Millis, A. J., Mackenzie, A. P. Magnetic Field-Tuned Quantum Criticality in the Metallic Ruthenate $Sr_3Ru_2O_7$, *Science* **294**, 329 (2001)

62. Khalifah, P., Osborn, R., Huang, Q., Zandbergen, H. W., Jin, R., Liu, Y., Mandrus, D., Cava, R. J. Orbital Ordering Transition in $La_4Ru_2O_{10}$, *Science* **297**, 2237 (2002)

63. Shi, Y., Guo, Y., Wang, X., Princep, A. J., Khalyavin, D., Manuel, P., Michiue, Y., Sato, A., Tsuda, K., Yu, S., Arai, M., Shirako, Y., Akaogi, M., Wang, N., Yamaura, K., Boothroyd, A. T. A Ferroelectric-Like Structural Transition in a Metal, *Nature Materials* **12**, 1024 (2013)

64. Sergienko, I. A., Keppens, V., McGuire, M., Jin, R., He, J., Curnoe, S. H., Sales, B. C., Blaha, P., Singh, D. J., Schwarz, K., Mandrus, D. Metallic "Ferroelectricity" in the Pyrochlore $Cd_2Re_2O_7$, *Phys. Rev. Lett.* **92**, 065501 (2004)

65. Tachibana, M., Taira, N., Kawaji, H., Takayama-Muromachi, E. Thermal Properties of $Cd_2Re_2O_7$ and $Cd_2Nb_2O_7$ at the Structural Phase Transitions, *Phys. Rev. B* **82**, 054108 (2010)

66. Jin, R., He, J., McCall, S., Alexander, C. S., Drymiotis, F., Mandrus, D. Superconductivity in the Correlated Pyrochlore $Cd_2Re_2O_7$, *Phys. Rev. B* **64**, 180503(R) (2001)

67. Castellan, J. P., Gaulin, B. D., van Duijn, J., Lewis, M. J., Lumsden, M. D., Jin, R., He, J., Nagler, S. E., Mandrus, D. Structural Ordering and Symmetry Breaking in $Cd_2Re_2O_7$, *Phys. Rev. B* **66**, 134528 (2002)

68. Chaloupka, J., Jackeli, G., Khaliullin, G. Zigzag Magnetic Order in the Iridium Oxide Na_2IrO_3, *Phys. Rev. Lett.* **110**, 097204 (2013)

69. Bremholm, M., Dutton, S. E., Stephens, P. W., Cava, R. J. $NaIrO_3$—A Pentavalent Post-Perovskite, *Journal of Solid State Chemistry* **184**, 601 (2011)

70. Liu, X., Katukuri, V. M., Hozoi, L., Yin, W.-G., Dean, M. P. M., Upton, M. H., Kim, J., Casa, D., Said, A., Gog, T., Qi, T. F., Cao, G., Tsvelik, A. M., van den Brink, J., Hill, J. P. Testing the Validity of the Strong Spin-Orbit-Coupling Limit for Octahedrally Coordinated Iridate Compounds in a Model System Sr_3CuIrO_6, *Phys. Rev. Lett.* **109**, 157401 (2012)

71. Kim, C. H., Kim, H. S., Jeong, H., Jin, H., Yu, J. Topological Quantum Phase Transition in $5d$ Transition Metal Oxide Na_2IrO_3, *Phys. Rev. Lett.* **108**, 106401 (2012)

72. Bhattacharjee, S., Lee, S.-S., Kim, Y. B. Spin–Orbital Locking, Emergent Pseudo-Spin and Magnetic Order in Honeycomb Lattice Iridates, *New J. Phys.* **14**, 073015 (2012)

73. Mazin, I. I., Jeschke, H. O., Foyevtsova, K., Valentí, R., Khomskii, D. I. Na_2IrO_3 as a Molecular Orbital Crystal, *Phys. Rev. Lett.* **109**, 197201 (2012)

74. Singh, Y., Gegenwart, P. Antiferromagnetic Mott Insulating State in Single Crystals of the Honeycomb Lattice Material Na_2IrO_3, *Phys. Rev. B* **82**, 064412 (2010)

75. Liu, X., Berlijn, T., Yin, W.-G., Ku, W., Tsvelik, A., Kim, Y.-J., Gretarsson, H., Singh, Y., Gegenwart, P., Hill, J. P. Long-Range Magnetic Ordering in Na_2IrO_3, *Phys. Rev. B* **83**, 220403(R) (2011)

76. Choi, S. K., Coldea, R., Kolmogorov, A. N., Lancaster, T., Mazin, I. I., Blundell, S. J., Radaelli, P. G., Singh, Y., Gegenwart, P., Choi, K. R., Cheong, S.-W., Baker, P. J., Stock, C., Taylor, J. Spin Waves and Revised Crystal Structure of Honeycomb Iridate Na_2IrO_3, *Phys. Rev. Lett.* **108**, 127204 (2012)

77. Ye, F., Chi, S., Cao, H., Chakoumakos, B. C., Fernandez-Baca, J. A., Custelcean, R., Qi, T. F., Korneta, O. B., Cao, G. Direct Evidence of a Zigzag Spin-Chain Structure in the Honeycomb Lattice: A Neutron and X-ray Diffraction Investigation of Single-Crystal Na_2IrO_3, *Phys. Rev. B* **85**, 180403(R) (2012)

78. Singh, Y., Manni, S., Reuther, J., Berlijn, T., Thomale, R., Ku, W., Trebst, S., Gegenwart, P. Relevance of the Heisenberg-Kitaev Model for the Honeycomb Lattice Iridates A_2IrO_3, *Phys. Rev. Lett.* **108**, 127203 (2012)

79. Cao, G., Qi, T. F., Li, L., Terzic, J., Cao, V. S., Yuan, S. J., Tovar, M., Murthy, G., Kaul, R. K. Evolution of Magnetism in the Single-Crystal Honeycomb Iridates $(Na_{1-x}Li_x)_2IrO_3$, *Phys. Rev. B* **88**, 220414(R) (2013)

80. Cao, G., Qi, T. F., Li, L., Terzic, J., Yuan, S. J., DeLong, L. E., Murthy, G., Kaul, R. K. Novel Magnetism of Ir^{5+} ($5d^4$) Ions in the Double Perovskite Sr_2YIrO_6, *Phys. Rev. Lett.* **112**, 056402 (2014)

81. Tokura, Y., Kawasaki, M., Nagaosa, N. Emergent Functions of Quantum Materials, *Nature Physics* **13**, 1056 (2017)

82. Basov, D. N., Averitt, R. D., Hsieh, D. Towards Properties on Demand in Quantum Materials, *Nature Materials* **16**, 1077 (2017)
83. Alexander, C. S., Cao, G., Dobrosavljevic, V., McCall, S., Crow, J. E., Lochner, E., Guertin, R. P. Destruction of the Mott Insulating Ground State of Ca_2RuO_4 by a Structural Transition, *Phys. Rev. B* **60**, R8422(R) (1999)
84. Lin, X. N., Zhou, Z. X., Durairaj, V., Schlottmann, P., Cao, G. Colossal Magnetoresistance by Avoiding a Ferromagnetic State in the Mott System $Ca_3Ru_2O_7$, *Phys. Rev. Lett.* **95**, 017203 (2005)
85. Durairaj, V., Lin, X. N., Zhou, Z. X., Chikara, S., Ehami, E., Douglass, A., Schlottmann, P., Cao, G. Observation of Oscillatory Magnetoresistance Periodic in 1/B and B in $Ca_3Ru_2O_7$, *Phys. Rev. B* **73**, 054434 (2006)
86. Crawford, M. K., Harlow, R. L., Marshall, W., Li, Z., Cao, G., Lindstrom, R. L., Huang, Q., Lynn, J. W. Structure and Magnetism of Single Crystal $Sr_4Ru_3O_{10}$: A ferromagnetic Triple-Layer Ruthenate, *Phys. Rev. B* **65**, 214412 (2002)
87. Cao, G., McCall, S., Shepard, M., Crow, J. E., Guertin, R. P. Thermal, Magnetic, and Transport Properties of Single-Crystal $Sr_{1-x}Ca_xRuO_3$ ($0 \leq x \leq 1.0$), *Phys. Rev. B* **56**, 321 (1997)
88. Cao, G., McCall, S., Freibert, F., Shepard, M., Henning, P., Crow, J. E. Observation of an Anomalous Quasi-One-Dimensional Behavior in $Na_2Ru_4O_{9-\delta}$ Single Crystals, *Phys. Rev. B* **53**, 12215 (1996)
89. Wang, J. C., Terzic, J., Qi, T. F., Ye, F., Yuan, S. J., Aswartham, S., Streltsov, S. V., Khomskii, D. I., Kaul, R. K., Cao, G. Lattice-Tuned Magnetism of $Ru^{4+}(4d^4)$ ions in Single Crystals of the Layered Honeycomb Ruthenates Li_2RuO_3 and Na_2RuO_3, *Phys. Rev. B* **90**, 161110(R) (2014)
90. Cao, G., Xin, Y., Alexander, C. S., Crow, J. E. Weak Ferromagnetism and Spin-Charge Coupling in Single-Crystal Sr_2YRuO_6, *Phys. Rev. B* **63**, 184432 (2001)
91. Perry, R. S., Baumberger, F., Balicas, L., Kikugawa, N., Ingle, N. J. C., Rost, A., Mercure, J. F., Maeno, Y., Shen, Z. X., Mackenzie, A. P. Sr_2RhO_4: a New, Clean Correlated Electron Metal, *New J. Phys.* **8**, 175 (2006)
92. Cao, G., Durairaj, V., Chikara, S., Parkin, S., Schlottmann, P. Partial Antiferromagnetism in Spin-Chain $Sr_5Rh_4O_{12}$, $Ca_5Ir_3O_{12}$, and Ca_4IrO_6 Single Crystals, *Phys. Rev. B* **75**, 134402 (2007)
93. Luo, Y., Cao, C., Si, B., Li, Y., Bao, J., Guo, H., Yang, X., Shen, C., Feng, C., Dai, J., Cao, G., Xu, Z. Li_2RhO_3: A Spin-Glassy Relativistic Mott Insulator, *Phys. Rev. B* **87**, 161121(R) (2013)
94. Terzic, J., Zheng, H., Ye, F., Zhao, H. D., Schlottmann, P., De Long, L. E., Yuan, S. J., Cao, G. Evidence for a Low-Temperature Magnetic Ground State in Double-Perovskite Iridates with $Ir^{5+}(5d^4)$ ions, *Phys. Rev. B* **96**, 064436 (2017)
95. Zhao, H., Ye, F., Zheng, H., Hu, B., Ni, Y., Zhang, Y., Kimchi, I., Cao, G. Ground State in the Novel Dimer Iridate $Ba_{13}Ir_6O_{30}$ with $Ir^{6+}(5d^3)$ Ions, *Phys. Rev. B* **100**, 064418 (2019)
96. Terzic, J., Wang, J. C., Ye, F., Song, W. H., Yuan, S. J., Aswartham, S., DeLong, L. E., Streltsov, S. V., Khomskii, D. I., Cao, G. Coexisting Charge and Magnetic Orders in the Dimer-Chain Iridate $Ba_5AlIr_2O_{11}$, *Phys. Rev. B* **91**, 235147 (2015)
97. Okamoto, Y., Nohara, M., Aruga-Katori, H., Takagi, H. Spin-Liquid State in the S=1/2 Hyperkagome Antiferromagnet $Na_4Ir_3O_8$, *Phys. Rev. Lett.* **99**, 137207 (2007)
98. Cao, G., Durairaj, V., Chikara, S., DeLong, L. E., Parkin, S., Schlottmann, P. Non-Fermi-Liquid Behavior in Nearly Ferromagnetic $SrIrO_3$ Single Crystals, *Phys. Rev. B* **76**, 100402(R) (2007)

99. Yanagishima, D., Maeno, Y. Metal-Nonmetal Changeover in Pyrochlore Iridates, *Journal of the Physical Society of Japan* **70**, 2880 (2001)

100. Qi, T. F., Korneta, O. B., Wan, X., DeLong, L. E., Schlottmann, P., Cao, G. Strong Magnetic Instability in Correlated Metallic $Bi_2Ir_2O_7$, *J. Phys.: Condens. Matter* **24**, 345601 (2012)

101. Baker, P. J., Möller, J. S., Pratt, F. L., Hayes, W., Blundell, S. J., Lancaster, T., Qi, T. F., Cao, G. Weak Magnetic Transitions in Pyrochlore $Bi_2Ir_2O_7$, *Phys. Rev. B* **87**, 180409(R) (2013)

102. Shinaoka, H., Miyake, T., Ishibashi, S. Noncollinear Magnetism and Spin-Orbit Coupling in $5d$ Pyrochlore Oxide $Cd_2Os_2O_7$, *Phys. Rev. Lett.* **108**, 247204 (2012)

103. Taguchi, Y., Oohara, Y., Yoshizawa, H., Nagaosa, N., Tokura, Y. Spin Chirality, Berry Phase, and Anomalous Hall Effect in a Frustrated Ferromagnet, *Science* **291**, 2573 (2001)

104. Erickson, A. S., Misra, S., Miller, G. J., Gupta, R. R., Schlesinger, Z., Harrison, W. A., Kim, J. M., Fisher, I. R. Ferromagnetism in the Mott Insulator Ba_2NaOsO_6, *Phys. Rev. Lett.* **99**, 016404 (2007)

105. Shi, Y. G., Guo, Y. F., Yu, S., Arai, M., Belik, A. A., Sato, A., Yamaura, K., Takayama-Muromachi, E., Tian, H. F., Yang, H. X., Li, J. Q., Varga, T., Mitchell, J. F., Okamoto, S. Continuous Metal-Insulator Transition of the Antiferromagnetic Perovskite $NaOsO_3$, *Phys. Rev. B* **80**, 161104(R) (2009)

106. Yonezawa, S., Muraoka, Y., Matsushita, Y., Hiroi, Z. Superconductivity in a Pyrochlore-Related Oxide KOs_2O_6, *J. Phys.: Condens. Matter* **16**, L9 (2004)

107. Hasan, M. Z., Kane, C. L. Colloquium: Topological Insulators, *Rev. Mod. Phys.* **82**, 3045 (2010)

108. Yan, B., Felser, C. Topological Materials: Weyl Semimetals, *Annual Review of Condensed Matter Physics* **8**, 337 (2017)

Part 2

Novel Phenomena in 4*d*- and 5*d*-Transition Metal Oxides

Chapter 2

Spin-Orbit Interactions in Ruddlesden-Popper Phases $Sr_{n+1}Ir_nO_{3n+1}$ (n = 1, 2, and ∞)

2.1 Overview

Iridium features a high atomic number, $Z = 77$, and is one of the nine least-abundant elements in the Earth's crust, but relatively common in meteorites. Iridium oxides or iridates have drawn growing attention in recent years due chiefly to the distinct influence of strong spin-orbit interactions (SOI) on their physical properties. Traditional arguments suggest that iridates should be more metallic and less magnetic than materials based upon 3d- and 4f-elements, because 5d-electron orbitals are more extended in space, which increases their electronic bandwidth (see Section 1.6). This conventional wisdom conflicts with early observations of two empirical trends inherent in iridates such as the Ruddlesden-Popper phases, $Sr_{n+1}Ir_nO_{3n+1}$ (n = 1 and 2; n defines the number of Ir-O layers in a unit cell) [1–5] and hexagonal perovskite $BaIrO_3$ [6]. First, complex magnetic states occur with high critical temperatures but unusually low ordered moments. Second, "exotic" insulating states, rather than metallic states, are commonly observed [1–6] (see Chapter 1, Table 1.3).

The early observations during the 1990s and early 2000s [1–6] signaled new physics unique to the 5d-electron-based materials and motivated extensive investigations in recent years, which eventually led to the recognition that a rare interplay of on-site Coulomb repulsion, U, crystalline fields, and strong SOI has unique, intriguing consequences in iridates. The most profound manifestation of such an interplay is characterized by the $J_{eff} = 1/2$ Mott state in Sr_2IrO_4, which was first identified in 2008 [7–9]. This quantum state represents novel physics and has since generated a surge of interest in this class of materials [10].

It is now recognized that the strong SOI can drive novel narrow-gap Mott states in iridates. As discussed in Chapter 1, Section 1.6, the SOI is a relativistic effect that scales with Z^2 (Z is the atomic number; e.g., Z = 29 and 77 for Cu and Ir, respectively). It is approximately 0.4 eV in iridates (compared to ~ 20 meV in 3d-materials; see Table 1.2), and splits the otherwise broad t_{2g} bands (**Fig. 2.1a**) into states with $J_{eff} = 1/2$ and $J_{eff} = 3/2$,

Physics of Spin-Orbit-Coupled Oxides. Gang Cao and Lance E. DeLong, Oxford University Press (2021). © Gang Cao and Lance E. DeLong.
DOI: 10.1093/oso/9780199602025.003.0002

the latter having lower energy [7] (**Fig. 2.1b**). Since Ir^{4+} (5d^5) ions provide five 5d-electrons to bonding states, four of them fill the lower J$_{eff}$ = 3/2 bands, and one electron partially fills the J$_{eff}$ = 1/2 band, W, where the Fermi level E$_F$ resides. The J$_{eff}$ = 1/2 band W is so narrow that even a reduced U (~ 0.5 eV) due to the extended nature of 5d-electron orbitals is sufficient to open an energy gap Δ supporting the insulating state in the iridates (**Fig. 2.1b**) [7]. Note that the ratio of U/W is the critical localization parameter. As discussed in Chapter 1, the splitting between the J$_{eff}$ = 1/2 and J$_{eff}$ = 3/2 bands narrows as the dimensionality (i.e., n) increases in Sr$_{n+1}$Ir$_n$O$_{3n+1}$, and the two bands progressively broaden and contribute to the density of states near the Fermi surface. In particular, the bandwidth W of the J$_{eff}$ = 1/2 band increases from 0.48 eV for n = 1 to 0.56 eV for n = 2 and 1.01 eV for n = ∞ [8]. The ground state evolves with decreasing charge gap Δ, from a robust insulating state for Sr$_2$IrO$_4$ (n = 1) to a metallic state for SrIrO$_3$ (n = ∞) as n increases (see Fig1.9).

The SOI-driven J$_{eff}$ = 1/2 model captures the essence of physics of many iridates with tetravalent Ir^{4+}(5d^5) ions. However, it is a single-particle approach that is limited to situations where Hund's rule interactions among the electrons can be neglected, and it may break down when non-cubic crystal fields, which are not taken into account in this model, become comparable to the SOI. This may result in a large overlap of J$_{eff}$ = 1/2 and J$_{eff}$ = 3/2 states, altering isotropic wavefunctions that the model is based upon. One such breakdown occurs in Sr$_3$CuIrO$_6$ due to strong non-octahedral crystal fields [11].

Nevertheless, a wide array of novel phenomena in iridates has been revealed in recent years (e.g., [10,12–18], and references therein). It has become apparent that materials with such a delicate interplay between SOI, U, and other competing interactions offer wide-ranging opportunities for the discovery of new physics and development of new devices, and it is not surprising that the physics of iridates is one of the most important topics in contemporary condensed matter physics. A great deal of theoretical work has appeared in response to early experiments on iridates, such as spin liquids in hyperkagome structures [19,20], superconductivity [21–27], Weyl semimetals with Fermi arcs, axion insulators [28], topological insulators, correlated topological insulators [16], Kitaev modes, 3D spin liquids with Fermionic spinons, topological semimetals

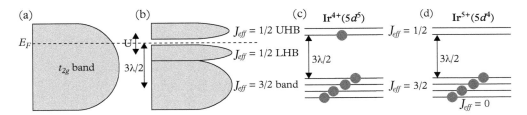

Fig. 2.1 *Band schematic: (a) the traditionally anticipated broad t$_{2g}$ band for 5d-electrons; (b) the splitting of the t$_{2g}$ band into J$_{eff}$ = 1/2 and J$_{eff}$ = 3/2 bands due to SOI; (c) Ir^{4+}(5d^5) ions provide five 5d-electrons; four of them fill the lower J$_{eff}$ = 3/2 bands, and one electron partially fills the J$_{eff}$ = 1/2 band where the Fermi level E$_F$ resides; (d) for Ir^{5+}(5d^4) ions, four 5d-electrons fill the J$_{eff}$ = 3/2 bands, leading to a singlet J$_{eff}$ = 0 state for the strong SOI limit [10].*

[12,16,29–37], etc. These and many other theoretical studies have advanced our understanding of 5d-based materials and motivated enormous activity in search of novel states in iridates. It is particularly intriguing that many proposals have met limited experimental confirmation thus far, which makes iridates even more interesting and challenging. Note that the SOI is a strong competitor with U and other interactions, which creates an entirely new hierarchy of energy scales (see Table 1.2); the lack of experimental confirmation underscores a critical role of subtle structural distortions that may dictate the low-energy Hamiltonian and need to be more thoroughly addressed both experimentally and theoretically.

There is a large number of existing iridates; and some of them are tabulated in **Table 2.1** (Note that Table 2.1 has some overlaps with Table 1.4). These iridates are grouped according to their underlying crystal structures. Exemplary phenomena along with valence states of the iridates are also listed.

Table 2.1 *Some Existing Iridates and Exemplary Phenomena*

Structure	Compound	Ir Ion	Exemplary Phenomena
Layered Perovskite	1. Sr_2IrO_4*	$Ir^{4+}(5d^5)$	1. J_{eff} = 1/2 insulator; T_N = 240 K [1– 4], S-shaped IV curves [4]*
	2. $Sr_3Ir_2O_7$*		2. J_{eff} = 1/2 or band insulator; T_N = 285 K [5], confined metal at 60 GPa [122]*
	3. $SrIrO_3$		3. Paramagnetic (semi-)metal (high-pressure phase) [8,123]
	4. Ba_2IrO_4		4. J_{eff} = 1/2 insulator; T_N ~240 K (high-pressure phase) [124]
	* Dopants		* Dopants: K [125], Ca [126], Mn [162], Ru [128,129], Rh [130], La [125], Eu [131], Tb [132]
Hexagonal Perovskite	1. $BaIrO_3$,	$Ir^{4+}(5d^5)$	1. J_{eff} = 1/2 Mott insulator; T_N = 183 K; CDW, S-shaped IV curves [6]
	2. $SrIrO_3$		2. Nearly ferromagnetic metal [133]
	3. $Ca_5Ir_3O_{12}$		3. Insulator; T_N = 10 K [134]
	4. Ca_4IrO_6		4. Insulator; T_N = 6 K [134]
	5. $Ba_3IrTi_2O_9$		5. Insulator; no long-range order [135]
	6. $Ba_3NdIr_2O_9$	$Ir^{4.5+}$	6. Magnetic insulator; T_N = 20 K [136]
	7. $Ba_3LiIr_2O_9$	$Ir^{4.5+}$	7. AFM order at T_N = 75 K [137]
	8. $Ba_3NaIr_2O_9$	$Ir^{4.5+}$	8. AFM order at T_N = 50 K [137]
Honeycomb	1. Na_2IrO_3	$Ir^{4+}(5d^5)$	1. Zigzag magnetic order; T_N = 18 K [29–32]
	2. α-Li_2IrO_3		2. Incommensurate order [138]
	3. β-Li_2IrO_3		3. Incommensurate order [139,140]
	4. γ-Li_2IrO_3		4. Incommensurate order [141,142]
	5. $Na_4Ir_3O_8$		5. Spin-liquid state [19]

(continued)

Table 2.1 *continued*

Structure	Compound	Ir Ion	Exemplary Phenomena
Pyrochlore	1. $Bi_2Ir_2O_7$ 2. $Pb_2Ir_2O_7$ 3. $RE_2Ir_2O_7$ RE = Rare Earth ion	$Ir^{4+}(5d^5)$	1 and 2: Metallic states; strong magnetic instability, etc. [143–145]. 3. Metal-insulator transition, insulating states, spin liquid, RE ionic size dependence [146,147]
Double Perovskite	1. Sr_2YIrO_6 2. Ba_2YIrO_6 3. Sr_2REIrO_6 4. Sr_2CoIrO_6 5. Sr_2FeIrO_6 6. La_2ZnIrO_6 7. La_2MgIrO_6 8. RE_2MIrO_6 M = Mg, Ni	$Ir^{5+}(5d^4)$ $Ir^{4+}(5d^5)$	1. Weak AFM state, correlated insulator [148] 2. Correlated insulator with magnetic moment [149,150] 3. Magnetic insulator [148] 4. Magnetic metal, T_{N1} = 60 K, T_{N2} = 120 K [151] 5. Magnetic insulator, T_N = 60 K [152] 6. Weak ferromagnetic insulator, T_N = 7.5 K [153] 7. Weak ferromagnetic insulator, T_N = 12 K [153] 8. Weak AFM ground state with canted spin [154]
Post-Perovskite	$NaIrO_3$ $CaIrO_3$	$Ir^{5+}(5d^4)$ $Ir^{4+}(5d^5)$	Paramagnetic insulator [155] Quasi-one-dimensional antiferromagnet [156,157]
Others	$Ba_5AlIr_2O_{11}$ Sr_3NiIrO_6 $Ba_4Ir_3O_{10}$ $Ba_{13}Ir_6O_{30}$	$Ir^{4.5+}$ $Ir^{4+}(5d^5)$ $Ir^{4+}(5d^5)$ $Ir^{6+}(5d^3)$	1. Spin-½ dimer chain, charge and magnetic orders [158] 2. One-dimensional chains [159] 3. New type of quantum liquid in unfrustrated lattice [160] 4. Novel magnetism with high effective J [161]

There are some remarkable empirical trends in the iridates, many of which are distinctly different from those of other materials:

- With a few exceptions (e.g., $SrIrO_3$, $Bi_2Ir_2O_7$), most iridates are antiferromagnetic (AFM) insulators (See Table 2.1), whose magnetic moments are merely a fraction of one Bohr magneton per Ir ion.
- The insulating state does not always closely track the magnetic state in iridates.
- All known iridates are inherently structurally distorted; a rotation of the IrO_6-octahedra about the c axis is commonplace but titling of the IrO_6 octahedra is rare in layered perovskite iridates.
- No first-order transitions are discerned thus far, and insulator-metal transitions tend to be gradual and continuous.

- The iridates do not metalize at high pressures; for example, Sr_2IrO_4 stays insulating up to 185 GPa.
- However, the physical properties of the iridates are extraordinarily sensitive to even slight chemical doping and/or slight lattice modifications. Doping-induced metallic states are common.
- However, the highly anticipated superconductivity remains elusive.

While some of these aspects are discussed in detail in the following sections, all these features are intimately associated with the lattice degrees of freedom, which play a central role that is hugely amplified by the strong SOI in iridates. There are also excellent review articles emphasizing theoretical aspects of iridates (e.g., [16,17]).

This chapter focuses on underlying properties of bulk single crystals of the Ruddlesden-Popper iridates, namely, Sr_2IrO_4 (Section 2.2), $Sr_3Ir_2O_7$ (Section 2.3), and Ir-deficient $SrIrO_3$ (Section 2.4). Their crystal structures are illustrated in **Fig. 2.2**. Each of the sections discusses the key features and representative properties of the corresponding compound.

Studies of thin films and heterostructures of iridates (e.g., [38–45]) are not included in this chapter. Since lattice properties are so critical to ground states of iridates, it is important to emphasize that epitaxial thin films with varied strain and/or heterostructures offer a unique, powerful tool for tuning ground states of iridates.

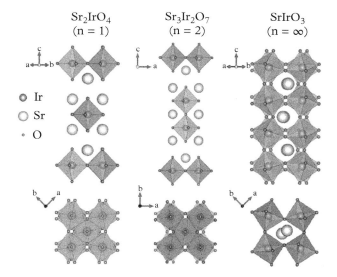

Fig. 2.2 *The crystal structure of $Sr_{n+1}Ir_nO_{3n+1}$ for n = 1, 2, and ∞. Note that the lower three panels illustrate the rotation of IrO_6 octahedra about the c axis for Sr_2IrO_4, $Sr_3Ir_2O_7$, and $SrIrO_3$, respectively.*

2.2 Novel Mott Insulator Sr₂IrO₄

Sr_2IrO_4 is the archetype $J_{eff} = 1/2$ Mott insulator with a Néel temperature $T_N = 240$ K [1–4,7,9], an energy gap $\Delta < 0.62$ eV (see **Fig. 2.3**) [46–48,127], and a relatively small magnetic coupling energy of 60–100 meV [23,47]. It is perhaps the most extensively studied iridate both experimentally and theoretically thus far, in part because of its distinct energy hierarchy featuring a strong SOI and its structural, electronic, and magnetic similarities to those of the cuprate La_2CuO_4, one hole per Ir or Cu ion, pseudospin- or spin-1/2 AFM, etc.

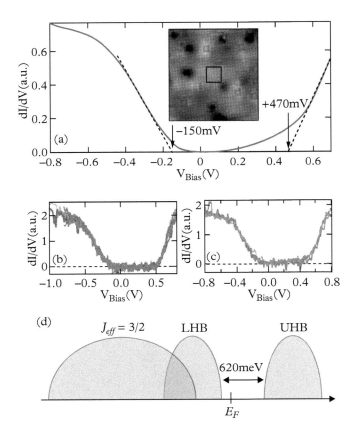

Fig. 2.3 *Sr_2IrO_4: The 620 meV energy gap. (a) The local density of states (LDOS) measured over the 2×2 nm² area indicated by the square in the image (inset). The dashed lines are drawn to indicate the band edges at −150 mV and +470 mV. (b,c) LDOS taken at locations away from defects, (b) without and (c) with the gradual increase. (d) Diagram showing energy bands with two important features: the 620 meV insulating gap and the overlap between the lower Hubbard band (LHB) and the $J_{eff} = 3/2$ band [48].*

2.2.1 Key Structural Features

A unique and important structural feature of Sr$_2$IrO$_4$, which has critical implications for the ground state, is a rotation of the IrO$_6$ octahedra about the c axis by ~11°, which results in a larger unit cell volume by a factor $\sqrt{2} \times \sqrt{2} \times 2$. It is commonly thought that Sr$_2$IrO$_4$ crystallizes in a tetragonal structure with space group $I4_1/acd$ (No. 142) with $a = b = 5.4846$ Å and $c = 25.804$ Å at 13 K [1–3]. More recently, studies of neutron diffraction (see Appendix, Section C) and second-harmonic generation (SHG) of single-crystal Sr$_2$IrO$_4$ [49,50] reveal structural distortions and forbidden reflections such as $(1, 0, 2n+1)$ for the space group $I4_1/acd$ over a wide temperature interval, 4 K < T < 600 K. Note that SHG or frequency doubling is an optical process where two photons with the same frequency interact with a material and create a new photon with twice the frequency of the initial photons (see Appendix, Section F). Nevertheless, these results indicate the absence of the c- and d-glide planes, leading to a further reduced structural symmetry with a space group $I4_1/a$ (No. 88) [49]. One defining characteristic of Sr$_2$IrO$_4$ and other iridates is that the strong SOI strongly couples physical properties to the lattice degrees of freedom [12,49–51,125], which is rare in other materials [52,53]. The rotation of IrO$_6$ octahedra, which corresponds to a distorted in-plane Ir1-O2-Ir1 bond angle, plays an extremely important role in determining the electronic and magnetic structures.

It is already recognized that the bond angle can be tuned via application of magnetic field [125], high pressure [54], electric field [51], epitaxial strain [38], and electrical current [55]. (Note that electrical-current-controlled quantum states is a new topic and is separately discussed in Chapter 5.) The lattice properties not only make the ground state readily tunable but also provide a new paradigm for development of functional materials and devices.

2.2.2 Magnetic Properties

Early experimental studies suggested that Sr$_2$IrO$_4$ was a weak ferromagnet with a Curie temperature at 240 K primarily because the temperature dependence of magnetization, along with a positive Curie-Weiss temperature, θ_{CW} = +236 K, appeared to be consistent with that of a weak ferromagnet (see **Fig. 2.4a**) [1–4]. It is now understood that the observed weak ferromagnetic behavior arises from an underlying canted AFM order [49]. This also raises a question as to why θ_{CW} (= +236 K), which is extrapolated from the inverse susceptibility $\Delta\chi^{-1}$ ($\Delta\chi = \chi(T) - \chi_o$, where χ_o is a temperature-independent contribution), is positive in the presence of the AFM ground state (**Fig. 2.4a**) [4,56,125]. Results of more recent neutron diffraction investigations of single-crystal Sr$_2$IrO$_4$ confirm that the system indeed undergoes a long-range AFM transition at 224(2) K with an ordered moment of 0.208(3) μ_B/Ir and a canted magnetic configuration within the basal plane [49]. The magnetic configuration illustrated in **Fig. 2.5** shows that magnetic moments are projected along the b axis with a staggered ↓↑↑↓ pattern along the c axis (**Fig. 2.5d**). Remarkably, these moments deviate 13(1)° away from the a axis (**Fig. 2.5c**), indicating that the magnetic moment canting rigidly tracks the staggered rotation of the IrO$_6$ octahedra discussed earlier. This strong coupling is a signature feature of the iridates

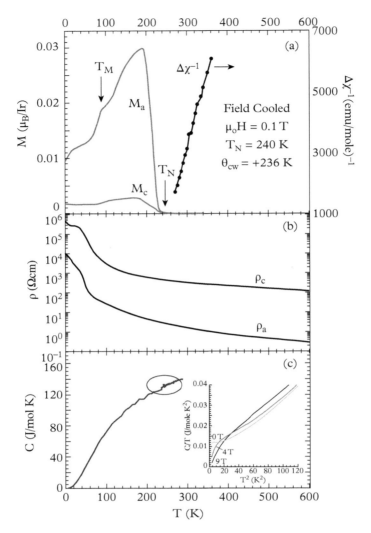

Fig. 2.4 *Sr_2IrO_4: (a) The temperature dependence of magnetization for the a and c axis, M_a and M_c, at 0.1 T and the inverse magnetic susceptibility $\Delta\chi^{-1}$(the right scale), (b) the electrical resistivity for the a- and c axis, ρ_a and ρ_c, and (c) the specific heat, C. Inset: C/T vs. T^2 at a few magnetic fields. T_M marks a magnetic anomaly near 100 K [125].*

in general and sharply contrasts with the situation in $3d$-oxides, such as $Ca_3Mn_2O_7$, where a collinear magnetic structure exists in a strongly distorted crystal structure [57]. Nevertheless, the magnetic canting results in 0.202(3) and 0.049(2) μ_B/Ir-site for the a axis and the b axis, respectively (**Fig. 2.5**) [49]. For comparison, the ordered moment extrapolated from the magnetization, which measures average magnetic moment, is less than 0.08 μ_B/Ir within the basal plane [56].

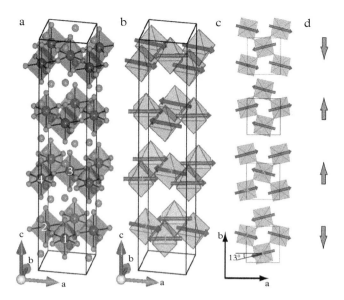

Fig. 2.5 *Sr$_2$IrO$_4$: (a) The crystal structure of Sr$_2$IrO$_4$. Each IrO$_6$ octahedron rotates 11.8° about the c axis. The Ir atoms of the non-primitive basis are labeled 1, 2, 3, and 4 plus the body centering translation (1/2,1/2,1/2). (b) The refined magnetic structure from single-crystal neutron diffraction measurements. (c) The same magnetic moment configuration projected on the basal planes. (d) The net moment projected along the b axis for individual layers [49].*

A muon spin rotation (μSR; see Appendix, Section B) study reports a low-frequency mode that corresponds to the precession of weak ferromagnetic moments arising from a spin canting and a high-frequency mode resulting from the precession of the AFM sub-lattices [58]. Another study indicates a small energy gap for the AFM excitations, 0.83 meV, suggesting an isotropic Heisenberg dynamics [59]. Remarkably, a high-resolution inelastic light (Raman) scattering study of the low-energy magnetic excitation spectrum of Sr$_2$IrO$_4$ shows that the high-field (>1.5 T) in-plane spin dynamics is isotropic and governed by the interplay between the applied field and the small in-plane ferromagnetic spin components induced by the Dzyaloshinsky-Moriya interaction (see Section 1.7). However, the spin dynamics of Sr$_2$IrO$_4$ at lower fields (<1.5 T) exhibit important effects associated with interlayer coupling and in-plane anisotropy, including a spin-flop transition in Sr$_2$IrO$_4$ that occurs either discontinuously or via a continuous rotation of the spins, depending on the in-plane orientation of the applied field. These results show that in-plane anisotropy and interlayer coupling effects play important roles in the low-field magnetic and dynamical properties of Sr$_2$IrO$_4$ [131]. It is also found that Sr$_2$IrO$_4$ (as well as Sr$_3$Ir$_2$O$_7$) exhibits pronounced two-magnon Raman scattering features and that the SOI might not be strong enough to quench the orbital dynamics in the paramagnetic state [60].

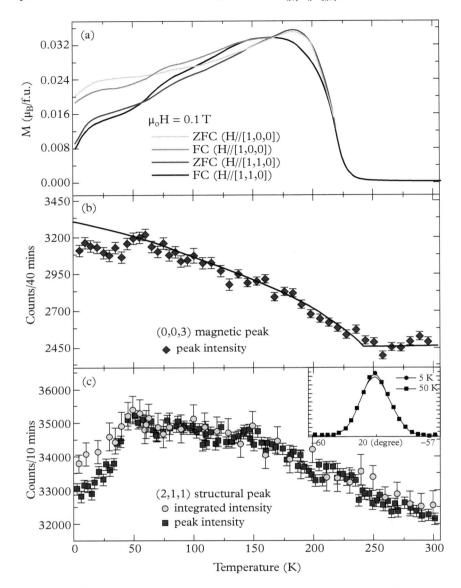

Fig. 2.6 *Sr_2IrO_4: Temperature dependence of (a) the basal-plane magnetization M measured in the zero-field-cooled (ZFC) and field-cooled (FC) sequences, with a magnetic field of 0.1 T applied along the [100] and [110] directions, respectively; (b) the peak intensity of the magnetic peak (0,0,3) that probes the canted magnetic moment (the solid line is a guide to the eye); and (c) the structural peak (2,1,1) associated with the in-plane rotation of the IrO_6 octahedra. The circles are the integrated intensities, and the squares are the peak intensities. Inset: The comparison of the θ-2θ scans at 5 K and 50 K [61].*

A close examination of the low-field magnetization, M(T), reveals two additional anomalies between 100 K and 25 K in M$_a$(T) and M$_c$(T) (see **Fig. 2.4a**), suggesting a moment reorientation. The existence of the reorientation is corroborated by the μSR study of Sr$_2$IrO$_4$ in which two structurally equivalent muon sites experience distinct local magnetic fields below 100 K characterized by the development of a second precession signal that is fully established below 20 K [58]. This behavior correlates with a change in the Ir1-O2-Ir1 bond angle that leads to the gradual magnetic moment reorientation below 100 K. This reorientation of the moments is at the root of the unusual magnetoresistivity [125] and giant magnetoelectric behavior that does not depend on the magnitude and spatial dependence of the magnetization, as conventionally anticipated [56].

Indeed, a neutron diffraction study uncovers a pronounced anomaly of the basal-plane IrO$_6$ rotation or the Ir1-O2-Ir1 bond angle below 50 K (**Fig. 2.6c**) [61] that closely tracks the pronounced anomalies observed in the electrical resistivity (**Fig. 2.4b**) and the magnetization (**Figs. 2.4a** and **2.6a**). The change in the Ir1-O2-Ir1 bond angle inevitably causes the magnetic moment orientation below 50 K because of the strong interlocking between the lattice and the magnetic moments, as discussed earlier. Remarkably, a hallmark of Sr$_2$IrO$_4$ is that there is no obvious correlation between the magnetic and transport properties near the Néel temperature at 240 K; more generally, there is an unusual (or the lack of) correlation between the magnetic and insulating states [10]. Such a direct correlation between the lattice and the physical properties illustrated in **Fig. 2.6** is thus particularly significant because the results reaffirm that it is the lattice degrees of freedom (rather than the spin degree of freedom) that drive the physical behavior in the iridate.

2.2.3 Transport Properties

The electrical resistivity of Sr$_2$IrO$_4$ for the *a* and *c* axis, ρ$_a$ and ρ$_c$, exhibits insulating behavior throughout the entire temperature range measured up to 600 K, as shown in **Fig. 2.4b**. The anisotropy, ρ$_c$/ρ$_a$, is significant, ranging from 10^2 to 10^3, although it is much smaller than 10^4–10^5 for La$_2$CuO$_4$ because of the extended nature of the 5d-electrons.

Both ρ$_a$ and ρ$_c$ exhibit an anomaly near 50 K, but not at T$_N$ (= 240 K), as shown in **Figs. 2.4a** and **2.4b**. As discussed earlier, this is a signature behavior of Sr$_2$IrO$_4$ in which transport properties exhibit no discernable anomaly corresponding to the AFM transition at T$_N$ = 240 K [4,56,125] but a pronounced anomaly near 50 K driven by the lattice anomaly (see **Fig. 2.6c**). In addition, the observed specific heat anomaly |ΔC| is tiny, ~4 mJ/mole K for Sr$_2$IrO$_4$ (**Fig. 2.4c**), in spite of its robust, long-range magnetic order at T$_N$ = 240 K. It is worth mentioning that the heat capacity C(T) below 10 K is predominantly proportional to T^3 at μ$_0$H = 0 and 9 T (**Fig. 2.4c**, inset), due to a Debye-phonon and/or magnon contributions from the AFM ground state. The field-shift [C(T,H)-C(T,0)]/C(T,0) ~ 16% at 9 T indicates a significant magnetic contribution to C(T), which is absent near T$_N$. The weak phase transition signatures suggest that thermal and transport properties may not be driven by the same mechanisms that dictate the magnetic behavior. Indeed, the energy gap for the AFM excitations, 0.83 meV [59], is almost negligible compared to the charge gap, 0.62 eV (**Fig. 2.3**). The effect of the magnetic state on the transport properties is thus inconsequential near T_N but more significant at low temperatures. This sharply contrasts with the behavior driven by strong couplings between the magnetic and charge gaps commonly observed in other correlated electron systems, particularly in $3d$-transition metal oxides [52,53].

In the following, we examine a few outstanding features of electrical and thermal transport properties that emphasize the importance of the lattice degrees of freedom.

Magnetoresistance

The electrical resistivity is coupled to the magnetic field in a peculiar fashion and so far no available model can describe the observed magnetoresistivity shown in **Fig. 2.7** [125]. We focus on a representative temperature T = 35 K. Both ρ_a (H||a) (**Fig. 2.7b**) and ρ_c(H||a) (**Fig. 2.7c**) exhibit an abrupt drop by ~60% near $\mu_oH = 0.3$ T applied along the a axis, where a metamagnetic transition occurs [4,56,63,125]. These data partially track the field dependences of M_a(H) and M_c(H) shown in **Fig. 2.7a**, and suggest a reduction of spin scattering. However, given the small ordered moment <0.08 μ_B/Ir, a reduction of spin scattering alone certainly cannot account for such a drastic reduction in ρ(H). Even more strikingly, for H||c axis, both ρ_a(H||c) and ρ_c(H||c) exhibit anomalies at $\mu_oH = 2$ T and 3 T, respectively, which lead to a large overall reduction of resistivity by more than 50%. However, no anomalies corresponding to these transitions in M_a(H) and M_c(H) are discerned. In addition, dM/dH shows no slope change near $\mu_oH = 2$ and 3 T. Such behavior is clearly not due to the Lorentz force, because ρ_c(H||c) exhibits the same behavior in a configuration where both the current and H are parallel to the c axis (**Fig. 2.7c**).

An essential contributor to conventional magnetoresistivity is spin-dependent scattering; negative magnetoresistance is often a result of the reduction of spin scattering due to spin alignment with increasing magnetic field. The data in **Fig. 2.7** therefore raise a fundamental question: why does the resistivity sensitively depend on the orientation of magnetic field H, but show no direct relevance to the measured magnetization when H||c axis? While no conclusive answers to the question are yet available, such varied magnetotransport behavior with temperature underscores the temperature-dependent Ir1-O2-Ir1 bond angle [49,58,61,125].

Moreover, such magnetotransport properties on the nanoscale are also examined using a point-contact technique [63]. Negative magnetoresistances up to 28% are discerned at modest magnetic fields (250 mT) applied within the basal-plane and electric currents flowing perpendicular to the plane. The angular dependence of the magnetoresistance shows a crossover from fourfold to twofold symmetry in response to an increasing magnetic field with angular variations in resistance from 1% to 14%, which is attributed to the crystalline component of anisotropic magnetoresistance and canted moments in the basal plane. The observed anisotropic magnetoresistance is large compared to that in $3d$-transition metal alloys or oxides (0.1%–0.5%) and is believed to be associated with the SOI [63].

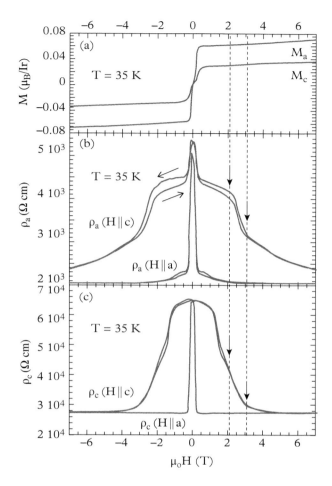

Fig. 2.7 *Sr$_2$IrO$_4$: The field dependences at T = 35 K of (a) the magnetizations M$_a$ and M$_c$. (b) The a-axis resistivity ρ$_a$ for H||a and H||c. (c) The c-axis resistivity ρ$_c$ for H||a and H||c [125].*

I-V Characteristic and Switching Effect

Early studies of iridates have also uncovered a distinct feature, namely, the non-Ohmic behavior [4,6]. The non-Ohmic behavior exhibits current-controlled negative differential resistivity (NDR) for both the *a*- and *c*-axis directions, as shown in **Fig. 2.8**. The I-V curve near the voltage threshold V$_{th}$ (onset of negative differential resistivity) for all temperatures shows a hysteresis effect. As the current I increases further (much higher than 100 mA in this case), the Ohmic behavior is restored and gives rise to an I-V curve characterized by an S shape. The S-shaped effect is categorically different from the more commonly

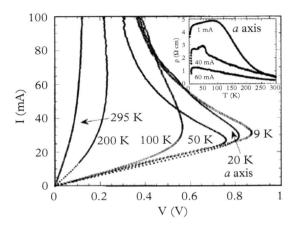

Fig. 2.8 *Sr_2IrO_4: Current I vs. voltage V for various temperatures. Inset: ρ along the a axis vs. temperature for various currents [4].*

seen N-shaped effect or the Gunn effect, which is referred to as voltage-controlled NDR, and attributed to electrons transferred between multienergy valleys [4,6]. It has been reported previously that the S-shaped effect is observed in some materials with a metal-insulator transition such as $CuIr_2S_{4-x}Se_x$, and is attributed to an electro-thermal effect [64]. A similar I-V characteristic has been found later in bulk single-crystal $BaIrO_3$ [6] and $Ca_3Ru_2O_7$ [65], and more recently in VO_2 [66]. The S-shaped I-V characteristic is restricted to the AFM insulating state. Its mechanism is still unclear, although it was suggested that the S-shaped effect might be related to a small band gap associated with charge density waves (CDW) [4,6,65]. In this case, the CDW is then pinned to the underlying lattice and slides relative to the lattice, giving rise to the NDR at $V > V_{th}$. Accordingly, one could assume a two-band model where the normal electrons and electrons in the CDW provide separate, independent channels for the conduction process [4,6,65]. It is clear that ρ is current-dependent and drastically decreases as I increases throughout the temperature range measured (see the inset of **Fig. 2.8**) [4]. A more recent study using nanoscale contacts reports a continuous reduction of the resistivity of Sr_2IrO_4 with increasing bias, which is characterized by a reduction in the transport activation energy by as much as 16% [51]. All this is a strong indication that applied electrical current can control both structural and physical properties via the strong SOI and magnetoelastic coupling. This is a key new research topic and is discussed in Chapter 5.

Thermal Conductivity via Pseudospin Transport

Pseudospin excitations give rise to significant thermal conductivity despite the insulating state (see **Fig. 2.9**) [67]. The analysis of the thermal conductivity reveals a relaxation of the pseudospin excitations at low temperatures. However, the relaxation rate dramatically increases as temperature rises due to thermally activated phonon scattering. The

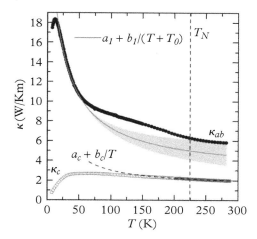

Fig. 2.9 *Sr$_2$IrO$_4$: The heat conductivity κ for the ab-plane (κ_{ab}) and along the c-direction (κ_c). Phononic fits to κ are shown as solid line. The fit to the ab-direction is given by the values $a_1 = 2.8\,W/Km$, $b_1 = 530\,W/m$, and $T_0 = 18K$. Similarly, the fit to the c-direction (dashed line) with $a_c = 1.43\,W/Km$, $b_c = 160\,W/m$ [67].*

comparison of the results with those for the cuprates with S = 1/2 spin excitations suggests a stronger coupling of the J_{eff} = 1/2 pseudospin excitations to the lattice [67]. This is consistent with the other experimental observations and the underlying characteristic of the iridates that physical properties are intimately associated with the lattice owing to the strong SOI. It is noteworthy that the anomaly near T_N is weak in κ_{ab} and absent in κ_c (**Fig. 2.9**).

2.2.4 Effects of High Pressure

A persistent insulating state at megabar pressures [68] is a unique, striking behavior of Sr$_2$IrO$_4$. In fact, all iridates studied so far — e.g., Sr$_3$Ir$_2$O$_7$, BaIrO$_3$, Na$_2$IrO$_3$, (Na$_{0.10}$Li$_{0.90}$)$_2$IrO$_3$ [69–75,122] — do not metalize at high pressures. This is in sharp contrast to the conventional wisdom that an insulating state will collapse in favor of an emergent metallic state at high pressures, because the unit cell volume inevitably shrinks and the average electron density must increase with pressure, thus the electronic bandwidth is expected to broaden and fill the insulating energy band gap [76,77]. One of the most dramatic results that supports traditional expectations is the discovered superconductivity in hydrogen sulfide above 200 K at megabar pressures [78].

The unusual response to pressure was observed in early studies in which Sr$_2$IrO$_4$ remains insulating up to 55 GPa [73] but the weak ferromagnetism vanishes near 20 GPa [54]. Such behavior is both puzzling and intriguing, and has thus motivated more investigations at higher pressures. A more recent high-pressure study of Sr$_2$IrO$_4$ using X-ray magnetic circular dichroism (XMCD; see Appendix, Section H) reveals a suppression of long-range AFM order at the range of 17–20 GPa, as shown in **Fig. 2.10** [79].

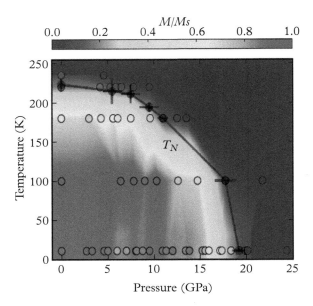

Fig. 2.10 Sr_2IrO_4: *The pressure-temperature evolution of magnetization at 0.5 T in neon pressure medium. Circles denote P = T values at which XMCD data were collected along isotherms. Color scale represents net magnetization normalized to saturation at P = 1 bar and T = 5.6 K. T_N at selected pressures, shown by solid black circles, was obtained by constraining M(T) to its functional form at ambient pressure [79].*

Interestingly, the persistence of strong exchange interactions characterized by a Curie-Weiss temperature of $\theta_{CW} = -209$ K up to 35 GPa suggests a highly frustrated magnetic state that prevents an otherwise strong tendency for a magnetic order (see **Fig. 2.11**). Note that without electronic correlations, a paramagnet would follow a Curie law, which is depicted with a dashed line in **Fig. 2.11**. Nevertheless, it is plausible that a novel quantum paramagnetic phase may exist at higher pressures in Sr_2IrO_4 [79].

Indeed, electrical resistance, synchrotron X-ray diffraction, and Raman scattering data over a much more extended range of pressures (up to 185 GPa) reveal a direct correlation between a structural phase transition from the native tetragonal $I4_1/acd$ phase to a lower-symmetry orthorhombic Pbca phase at a critical pressure $P_c = 40.6$ GPa, and a nearly concurrent strengthening of an insulating ground state of Sr_2IrO_4 that persists up to at least 185 GPa [68] (see Appendix, Section D for Raman scattering).

As shown in **Fig. 2.12**, the resistance R increases monotonically with decreasing temperature for pressure P = 0.6 GPa. With increasing P up to 32.1 GPa, R decreases by more than two orders of magnitude at low temperatures but retains insulating behavior (**Fig. 2.12**(a1)). This trend reverses for P > 32 GPa, as shown in **Fig. 2.12**(a2): R rapidly rises and almost fully recovers its initial, ambient-pressure value. A corresponding T-P contour plot in **Fig. 2.12** (b) illustrates a U-shaped pressure dependence of R with a minimum near 32 GPa.

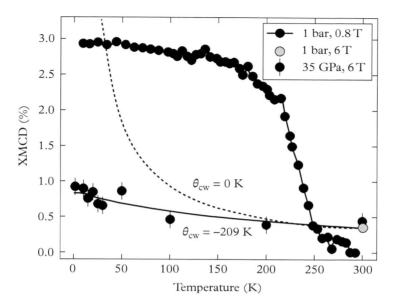

Fig. 2.11 *Sr₂IrO₄: The temperature dependence of XMCD signal showing both the ordered and disordered magnetic states at 1 bar and 35 GPa, respectively. The disordered state features a* $\theta_{cw} = -209$ K, *suggesting a strong tendency for a magnetic order. An anticipated Curie paramagnetic behavior (dashed line) is shown for comparison [79].*

However, below 80 K and at P >124 GPa, R decreases by more than 67%, compared to R at 81 GPa, and exhibits an apparent approach to saturation (**Fig. 2.12e**). The tendency toward saturation is both significant and intriguing, and may indicate a possible topological insulating state in which a saturated resistance at low temperatures could be a result of a pressure-induced surface state [44,45]. This result is interestingly relevant to a possible quantum paramagnetism or a topological state suggested earlier [79]. It is nevertheless clear that the electronic structure of Sr_2IrO_4 undergoes a significant change in the megabar range.

The rapid rise of R near 38 GPa is accompanied by a critical structural phase transition at $P_c = 40.6$ GPa, as shown in **Fig. 2.13** [68]. Nearly simultaneous changes in R and the crystal structure indicate a direct correlation between the retention of an insulating state and the structural distortions; that is, the persistent insulating state at megabar pressures is related to the significant reduction in symmetry incurred in the transition from the $I4_1/acd$ phase to the lower-symmetry *Pbca* phase. This structural change involves not only rotations, but also titling of IrO_6 octahedra at $P > P_c$. The striking stability of the insulating state over such a broad pressure interval of 38 to 185 GPa suggests two competing forces are at work: (1) There is a tendency for band broadening that must accompany a sizable volume compression and that favors metallic behavior. (2) There is a pressure-induced crystal distortion that generally weakens electron hopping and can

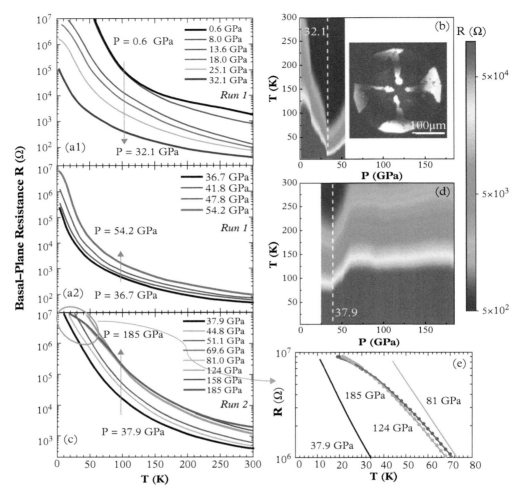

Fig. 2.12 Sr_2IrO_4: *The temperature dependence of the basal-plane resistance R over pressures ranging (a1) 0.6–32.1 GPa and (a2) 36.7–54.2 GPa for Run 1, and (c) 37.9–185 GPa for Run 2. Note that the gray arrows indicate the increase or decrease in R with increasing P. Corresponding contour plots are shown in (b) for the pressure range 0.6–54.2 GPa for Run 1, and in (d) for the pressure range 24.7–185 GPa for Run 2. The colors red and blue represent the highest and lowest resistance R, respectively; and other colors indicate intermediate resistance values. The white dashed lines mark the pressure regime of 32.1–37.9 where R reaches its minimum. (e) Data marked by the green oval circle in (c) is shown in an expanded plot of the near-saturated regime for R (i.e., P ≈ 185 GPa and T < 80 K). Inset in (b): A snapshot of the diamond anvil cell with a sample at 27 GPa [68].*

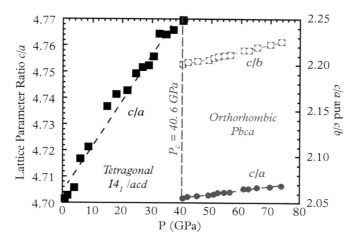

Fig. 2.13 *Sr₂IrO₄: The pressure dependence of the lattice parameter ratio of the c axis to the a axis, c/a, for the tetragonal phase, and c/a and c/b for the orthorhombic phase [68].*

lead to localization, which eventually prevails in the present case, given the recovered insulating state for P > 38 GPa.

The data in **Fig. 2.13** also show that the lattice c/a ratio increases significantly with rising pressure in both the tetragonal phase below P_c (= 40.6) and the orthorhombic phase above P_c. These observations may offer a reasonable explanation of the disappearance of weak ferromagnetism above 20 GPa, since the enhanced c/a ratio suggests that Sr₂IrO₄ becomes more two-dimensional, which is generally unfavorable for long-range magnetic order. In particular, the weak ferromagnetism is due to magnetic canting, which closely tracks the IrO₆ rotation. It is recognized that an elongation (compression) of the c axis weakens (enhances) the weak ferromagnetism, and facilitates either a collinear AFM or a paramagnetic state [12].

Nevertheless, Sr₂IrO₄ defies conventional Mott physics in that the insulating state and long-range AFM order do not always precisely accompany each other. The lattice symmetry and dynamics play a role in the formation of an insulating gap that is much more critical than traditionally recognized, and is illustrated in many studies in which the electronic ground state of the iridates critically hinges on the bond angle Ir1-O2-Ir1 of IrO₆ octahedra [80–82,125,130,132]. The unusual characteristic provides a new perspective for understanding the discrepancies between recent theoretical proposals and experimental results in iridates, including the absence of superconductivity in Sr₂IrO₄ discussed later. It has also helped revitalize discussions of Mott, Mott-Hubbard, and Slater insulators; in particular, the dependence of the charge gap formation on magnetic interactions in Sr₂IrO₄ [83,84,127]. Remarkably, a time-resolved optical study indicates that Sr₂IrO₄ is a unique system in which Slater- and Mott-Hubbard-type behaviors coexist [83], which might help explain the absence of anomalies at T_N in transport and thermodynamic measurements discussed earlier.

More generally, a persistent insulating state at megabar pressures raises an intriguing and fundamental issue: the strong exchange-correlation effects supported by a high, narrow peak in the density of states near the Fermi level may not lead to traditional (metallic/delocalized) Fermi liquid screening interactions in Sr_2IrO_4, as anticipated from Mott physics. We speculate that very large volume reductions and strong Coulomb correlations alternatively can stabilize highly directional bonds. This contrasts with Hartree-Fock mean-field theories that treat breaking of spherical symmetry by electron-electron interactions via spherical averaging of self-consistent Coulomb fields. The persistent insulating behavior in Sr_2IrO_4 indicates that the Hartree-Fock methodology is not well suited for treating situations where strong SOI and anisotropic correlations dictate that electron localization is dominant at very high densities.

2.2.5 Effects of Chemical Substitution

In stark contrast, a growing body of experimental evidence has shown that a metallic state can be readily realized via slight chemical doping, either electron (e.g., La doping [85,125]) or hole doping (e.g., K [125], Rh [130], or Ru [128,129] doping), for either Sr or Ir or oxygen, despite the sizable energy gap (~ 0.62 eV shown in **Fig. 2.3**). Electron doping adds extra electrons to the partially filled J_{eff} = 1/2 states, which is energetically favorable. Hole doping also adds additional charge carriers, resulting in a metallic state. However, because of the multi-orbital nature of the iridate, the mechanism of electron and hole doping in Sr_2IrO_4 may not be symmetrical. An important distinction lies in the energy gap to the nearest $5d$ states [17,127]. Nevertheless, the same effect of both electron and hole doping, as suggested by some experimental evidence, is to reduce the structural distortions or relax the buckling of IrO_6 octahedra, independent of the ionic radius of the dopant [125]. A dilute doping of either La^{3+} (electron doping) or K^+ (hole doping) ions for Sr^{2+} ions leads to a larger Ir1-O2-Ir1 bond angle θ, despite the considerable differences between the ionic radii of Sr, La, and K, which are 1.18 Å, 1.03 Å, and 1.38 Å, respectively. Empirical trends suggest that the reduced distortions, along with extra charge carriers, can effectively destabilize the J_{eff} = 1/2 state, leading to a metallic state, since hopping between active t_{2g} orbitals is critically linked to θ. Indeed, $ρ_a$ ($ρ_c$) is reduced by a factor of 10^8 (10^{10}) at low temperatures for mere x = 0.04 for La doping (see **Figs. 2.14a** and **2.14b**) [125]. Furthermore, for La doping of x = 0.04, there is a sharp downturn near 10 K, indicative of a rapid decrease in inelastic scattering [125], which is similar to an anomaly observed in oxygen-depleted or effectively electron-doped $Sr_2IrO_{4-δ}$ with δ = 0.04 [80]. Other studies of La-doped Sr_2IrO_4 show similar effects of La doping on physical properties, although behavior varies in detail [85]. Note that T_N decreases with La doping in $(Sr_{1-x}La_x)_2IrO_4$, and vanishes at x = 0.04, where the metallic state is fully established.

For Ru- [128,129] or Rh- [130] doped Sr_2IrO_4, effects of doping on magnetic and electronic properties are different in detail, but in essence Ru or Rh doping results in a metallic state that coexists with the AFM excitations. For example, for $Sr_2Ir_{1-x}Ru_xO_4$, the AFM excitations persist up to at least x = 0.77, and the maximum energy scale of the magnetic excitations at high dopings is comparable to that in undoped Sr_2IrO_4.

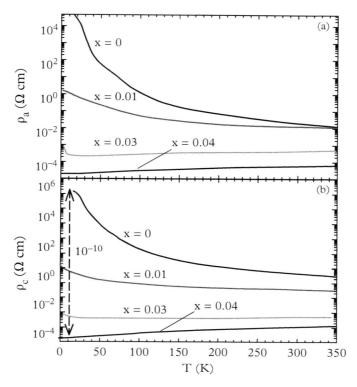

Fig. 2.14 *(Sr$_{1-x}$La$_x$)$_2$IrO$_4$ with 0 ≤ x ≤ 0.04: The temperature dependences of (a) the a-axis resistivity ρ$_a$, and (b) the c-axis resistivity ρ$_c$ [125].*

It is interesting that mere 3% Tb^{4+} substitution for Ir^{4+} effectively suppresses T$_N$ to zero but retains the insulating state, that is, the disappearance of the AFM state accompanies no emergence of a metallic state, as shown in **Fig. 2.15** [132].

Indeed, a neutron diffraction study confirms these changes [132]. As shown in **Fig. 2.16**, with increasing Tb concentration x, a signature magnetic peak at (1,0,2) for the AFM state becomes weakened at x = 0.005 (**Fig. 2.16a**) and eventually vanishes at x = 0.03 (**Fig. 2.16b**). The disappearance of the sharp magnetic peak associated with the canted antiferromagnetic configuration at x = 0 is accompanied by an emergent incommensurate magnetic order with wave vectors q_m = (0.95,0,0) and (0,0.95,0). The incommensurate magnetic order becomes better defined at x = 0.03 when the higher-temperature background is subtracted (**Fig. 2.16b**, inset). The intensity of the new peaks is much weaker compared to those at q = (1,0,2) for undoped Sr$_2$IrO$_4$. The new peaks exhibit a clear temperature dependence and evolve into a featureless background above 30 K. The pair of peaks at (0.95,0,0) and (0, 0.95, 0) suggest a possible spiral order with moments along the *c* axis or an incommensurate spin-density wave as neutron diffraction probes only the moment component perpendicular to the momentum transfer. It is likely that

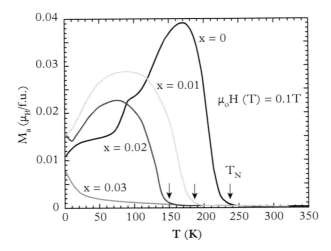

Fig. 2.15 $Sr_2Ir_{1-x}Tb_xO_4$: *The a-axis magnetization as a function of temperature at $\mu_oH = 0.1$ T. Note that mere 3% Tb doping readily suppresses the AFM order but the insulating state retains [132].*

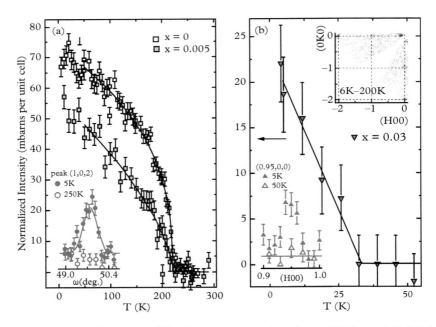

Fig. 2.16 $Sr_2Ir_{1-x}Tb_xO_4$: *The neutron diffraction: temperature dependence of (a) the peak (1,0,2) for x = 0 and x = 0.005. Inset: The rocking curve of the (1,0,2) peak for x = 0.005 at 5 K and 250 K. (b) Peaks (0.95,0,0) and (0,0.95,0) for x = 0.03 in the same units as in (a). Upper inset: The HK0-slice from time-of-flight neutron data. Lower inset: High-resolution triple-axis results of (H,0,0) scan at 5 K and 50 K. The peaks are incommensurate at (0.95,0,0) and (0,0.95,0) [132].*

the magnetic moment of Tb ions, which tends to polarize the magnetic moment of surrounding Ir ions along with it, is ferromagnetically aligned along the *c* axis or forms magnetic polarons. Generally, a *c*-axis alignment is more energetically favorable when the tetragonal crystal field effect (CFE) is enhanced [12]. A theoretical study suggests that the interaction between the magnetic moments on the impurity Tb^{4+} ion and its surrounding Ir^{4+} ions can be described by an Ising-like interaction that favors the magnetic moments across each bond to align along the bond direction. This interaction quenches magnetic vortices near the impurities and drives a reentrant transition out of the AFM phase, leading to a complete suppression of the Néel temperature [86].

A systematic structural, transport, and magnetic study of Ca- or Ba-doped Sr$_2$IrO$_4$ single crystals reveals that isoelectronically substituting Ca^{2+} or Ba^{2+} ion for the Sr^{2+} ion, which provides no additional charge carriers, causes no change in the Néel temperature, which remains at 240 K (**Fig. 2.17a**), but drastically reduces the electrical resistivity by up to five orders of magnitude, or even precipitates a sharp insulator-to-metal transition at lower temperatures (**Fig. 2.17b**); that is, the vanishing insulating state accompanies an essentially unchanged Néel temperature in (Sr$_{1-x}$A$_x$)$_2$IrO$_4$ where A = Ca or Ba [87]. This observation brings to light an intriguing difference between chemical pressure and

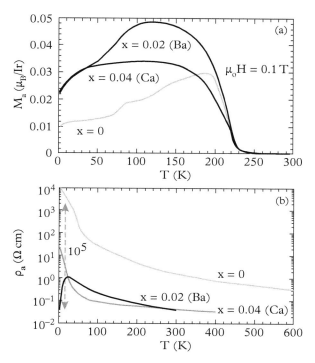

Fig. 2.17 *(Sr$_{1-x}$A$_x$)$_2$IrO$_4$ with A = Ba or Ca: The temperature dependence of (a) the a-axis magnetization M$_a$ and (b) the a-axis resistivity ρ$_a$ for x = 0, x = 0.02 (Ba) and x = 0.04 (Ca). Note that T$_N$ remains essentially unchanged but the magnitude of ρ$_a$ drops by up to five orders of magnitude.*

applied pressure, the latter of which does suppress the long-range magnetic order in Sr_2IrO_4. This difference reveals the importance of the Ir1-O2-Ir1 bond angle and homogenous volume compression in determining the magnetic ground state.

All results underscore that the magnetic transition plays a nonessential role in the formation of the charge gap in the spin-orbit-tuned iridate [87]. Indeed, more recent studies indicate that the antiferromagnetic and insulating states coexist in a peculiar manner in Sr_2IrO_4 [163].

2.2.6 Elusive Superconductivity and Odd-Parity Hidden Order

Sr_2IrO_4 bears key structural, electronic, and magnetic features similar to those of La_2CuO_4 [1–4,7,9,12,23], as pointed out in the beginning of this section. In particular, the magnon dispersion in Sr_2IrO_4 is well described by an AFM Heisenberg model with an effective spin one-half on a square lattice, which is similar to that in the cuprate. The magnon bandwidth in Sr_2IrO_4 is of 200 meV, as compared to ~300 meV in La_2CuO_4 [1–4,7,9,23]. This smaller bandwidth is consistent with other energy scales, such as hopping t and U, which are uniformly smaller by approximately 50% in Sr_2IrO_4 than in the cuprate [21]. Largely because of these apparent similarities, a pseudospin-singlet d-wave superconducting phase with a critical temperature, T_c, approximately half of that in the superconducting cuprate is anticipated in electron-doped Sr_2IrO_4, whereas pseudospin triplet pairing may emerge in the hole-doped iridate so long as the Hund's rule coupling is not strong enough to drive the ground state to a ferromagnetic state [17,21–27]. A growing list of theoretical proposals has motivated extensive investigations both theoretically and experimentally in search of superconductivity in the iridates in recent years. Indeed, there is some experimental evidence signaling behavior parallel to that of the cuprates. Besides results of resonant inelastic X-ray scattering (RIXS; see Appendix, Section E) that indicate similar magnon dispersion in Sr_2IrO_4 to that in La_2CuO_4 [23,88], Raman studies also reveal excitations at 0.7 eV in Sr_2IrO_4, about 50% smaller than 1.5 eV observed in La_2CuO_4 [89]. Studies of angle-resolved photoemission spectroscopy (ARPES; Appendix, Section A) reveal an evolution of Fermi surface with doping (e.g., pseudogaps, Fermi arcs) that is strikingly similar to that in the cuprates, and is observed in Sr_2IrO_4 with either electron doping (e.g., La doping) or hole doping (e.g., Rh doping) [84,90]. For example, the Fermi surface segments and pseudogaps for Rh-doped Sr_2IrO_4 (hole doing) are illustrated in **Fig. 2.18**. It is particularly interesting that a temperature and doping dependence of Fermi arcs at low temperatures is observed with in-situ K doping in the cleaved crystal surface of Sr_2IrO_4. This phenomenology is strikingly similar to that of the cuprates [90]. The similarities to the cuprates are further signified in another RIXS study of La-doped Sr_2IrO_4 [91]. This study uncovers well-defined dispersive magnetic excitations. The dispersion is almost intact along the anti-nodal direction, but exhibits significant softening along the nodal direction, similar to those in hole-doped cuprates [91].

However, superconductivity, which is characterized by zero-resistivity and diamagnetism, remains markedly elusive, although a metallic state is a common occurrence in doped Sr_2IrO_4. The absence of superconductivity may be a manifestation of the particular importance of lattice properties that separate the iridates from the cuprates, in spite

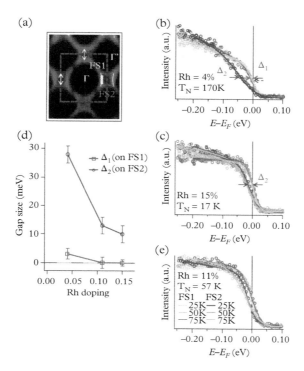

Fig. 2.18 *Sr$_2$Ir$_{1-x}$Rh$_x$O$_4$: Fermi surface (FS) segments and pseudogaps: (a) the FS spectral weight for x = 15%, with a hole-like Fermi pocket centered around the (π, 0) point of the unfolded (blue dashed) Brillouin zone. The FS pocket is separated into segments FS1 (yellow) and FS2 (blue), with FS1 facing Γ and FS2 facing Γ'. Q vectors (white arrows) are possible density wave nesting vectors. (b,c) Energy distribution curves (EDCs) from multiple locations along the FS1 and FS2 segments (yellow and blue, respectively) taken from x = 4% and x = 15%. The leading edges of most EDCs do not reach E$_F$, suggesting an occurrence of pseudogaps. Gap sizes are shown in (b) and (c), and compiled in (d), with Δ$_1$ labeling the gaps from FS1 and Δ$_2$ the gaps from FS2. (e) EDCs from FS1 (dashed) and FS2 (solid) showing minimal temperature dependence across the magnetic phase transition for x = 11% [84].*

of all the similarities between these two classes of materials described herein and in the literature. Indeed, the lattice-dependence of physical properties is much weaker in the cuprates, where the SOI is generally negligible. This point is increasingly recognized both experimentally [10] and theoretically [92]. Given the extraordinary sensitivity to structural distortions in the iridates, realizing superconductivity and other theoretically predicted states in these materials clearly demands innovative experimental approaches. Recently, a field-altering technology proposed to directly address these materials challenges shows promising results [93], and is discussed in Chapter 6.

Interestingly, a SHG study of single-crystal Sr$_2$Ir$_{1-x}$Rh$_x$O$_4$ has revealed an odd-parity hidden order that sets in at a temperature T$_\Omega$, emerging prior to the formation of the AFM state, as shown in **Fig. 2.19** [50,94]. This order breaks both the spatial inversion

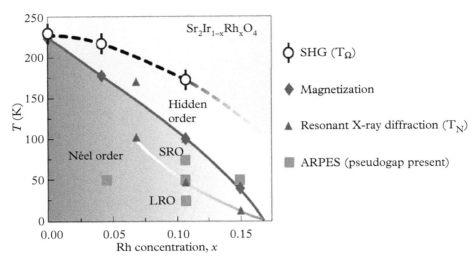

Fig. 2.19 *The phase diagram of temperature vs. Rh doping for $Sr_2Ir_{1-x}Rh_xO_4$. Note the boundaries of the hidden order and the long-range order (LRO) and short-range order (SRO) in the AFM regions [50].*

and rotational symmetries of the underlying tetragonal lattice, which is expected from an electronic phase that has the symmetries of a magnetoelectric loop-current order [95,96]. The onset temperature of this phase is monotonically suppressed with hole doping, although much more weakly than is the Néel temperature, revealing an extended region of the phase diagram with purely hidden order (**Fig. 2.19**). A more recent study suggests that the loop current flows along a diagonal direction in the IrO_4 square [164].

2.3 Borderline Insulator: $Sr_3Ir_2O_7$

The splitting between the $J_{eff} = 1/2$ and $J_{eff} = 3/2$ bands narrows as the effective dimensionality increases in $Sr_{n+1}Ir_nO_{3n+1}$, and the two bands progressively broaden and contribute a finite density of states near the Fermi surface. In particular, the bandwidth W of the $J_{eff} = 1/2$ band increases from 0.48 eV for n = 1 to 0.56 eV for n = 2 and 1.01 eV for n = ∞ (see **Fig. 1.9**) [8,46]. The ground state evolves with decreasing charge gap Δ as n increases from a robust insulating state for Sr_2IrO_4 (n = 1) to a metallic state for $SrIrO_3$ (n = ∞). A well-defined yet weak insulating state exists in the case of $Sr_3Ir_2O_7$ (n = 2) [5]. Given the delicate balance between relevant interactions, $Sr_3Ir_2O_7$ is theoretically predicted to be at the border between a collinear AFM insulator and a spin-orbit Mott insulator [17,97]. Early experimental observations indicated an insulating state and a long-range magnetic order at $T_N = 285$ K with an unusual magnetization reversal below 50 K (**Fig. 2.20**) [5]. The value of the charge gap (180 meV) [46] is much smaller than that of Sr_2IrO_4 (0.62 eV), but the magnetic gap is unusually large at 92 meV [98]. The borderline nature of the weak insulating state of $Sr_3Ir_2O_7$ is apparent in an ARPES study of the near-surface electronic structure, which exhibits weak metallicity evidenced by finite electronic spectral weight at the Fermi level [99].

2.3.1 Key Structural Features

$Sr_3Ir_2O_7$ has strongly coupled, double Ir-O layers separated from adjacent double layers along the c axis by Sr-O interlayers, as shown in **Fig. 2.2**. The crystal structure features an orthorhombic cell with $a = 5.5221$ Å, $b = 5.5214$ Å, $c = 20.9174$ Å, and *Bbca* symmetry [5] (a more recent study suggests a space group *C2/c* [100]). The IrO_6 octahedra are elongated along the crystallographic c axis. The average Ir-O apical bond distances along the c axis are 2.035 Å and 1.989 Å in the IrO_6 octahedra in the double layers. Like those in Sr_2IrO_4, the IrO_6 octahedra are rotated about the c axis by 11° at room temperature. It is found that within a layer, the rotations of the IrO_6 octahedra alternate in sign, forming a staggered structure in which the two layers are out of phase with each other [5].

2.3.2 Magnetic Properties

The onset of magnetic order is observed at $T_N = 285$ K in $Sr_3Ir_2O_7$ (**Fig. 2.20**) [5]. It is generally recognized that the magnetic ground state is AFM and closely associated with the rotation of the IrO_6 octahedra about the c axis, which characterizes the crystal structure of both Sr_2IrO_4 and $Sr_3Ir_2O_7$ [1–5,100]. Indeed, the temperature dependence of the magnetization M(T) closely tracks the rotation of the octahedra, as reflected in the

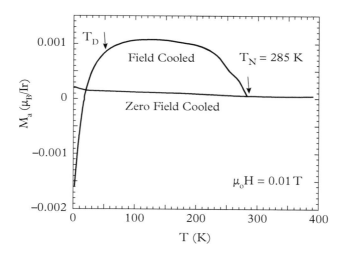

Fig. 2.20 *$Sr_3Ir_2O_7$: The temperature dependence of the magnetization for the a-axis M_a. The AFM order occurs at $T_N = 285$ K and the magnetization reversal starts below $T_D = 50$ K. Note that M_a is measured via both field-cooled and zero-field-cooled sequences. No magnetic order can be discerned in the zero-field-cooled sequence [5].*

Ir-O-Ir bond angle θ [10]. Unlike Sr_2IrO_4, $Sr_3Ir_2O_7$ exhibits an intriguing magnetization reversal in the *a*-axis magnetization $M_a(T)$ below $T_D = 50$ K; both T_N and T_D can be observed only when the system is field-cooled (FC) from above T_N. This magnetic behavior is robust but not observed in the zero-field-cooled (ZFC) magnetization, which instead remains positive and displays no anomalies that are seen in the FC magnetization (**Fig. 2.20**) [5]. That all magnetic anomalies are absent in the case of ZFC measurements signals that magnetostriction may occur near T_N and "lock up" a certain magnetic configuration.

This magnetic behavior of $Sr_3Ir_2O_7$ is distinctively different from that of Sr_2IrO_4. A number of experimental and theoretical studies indicate that magnetic moments are aligned along the *c* axis (see **Fig. 2.21**) [101,102], but there may exist a nearly degenerate magnetic state with canted spins in the basal plane [103]. A transition to a collinear antiferromagnet via multi-orbital Hubbard interactions is predicted within the mean-field approximation [17,103]. Indeed, a resonant X-ray diffraction study reveals an easy collinear antiferromagnetic structure along the *c* axis in $Sr_3Ir_2O_7$ (see **Fig. 2.21**) rather than within the basal plane, as observed in Sr_2IrO_4 (**Fig. 2.5**). This study further suggests a spin-flop transition as a function of the number of IrO_2 layers, due to strong competition among intra- and interlayer bond-directional, pseudodipolar interactions [102].

The magnetic configuration of $Sr_3Ir_2O_7$ is highly sensitive to the lattice structure. In particular, the staggered rotation of IrO_6 octahedra between adjacent layers plays a crucial role in both insulating and magnetic states [17,103]. $Sr_3Ir_2O_7$ is a unique magnetic insulator, given its tiny magnetic moment [5], and is more prone to undergo a transition

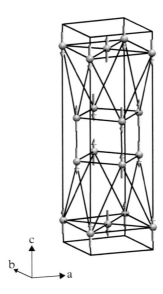

Fig. 2.21 *$Sr_3Ir_2O_7$: The magnetic structure indicates that all magnetic moments are aligned along the c axis [101], in sharp contrast to that for Sr_2IrO_4.*

[17,103]. It is not surprising that any slight perturbation such as magnetic field could induce a canted spin structure, as indicated in a magnetic X-ray scattering study [104]. In addition, a study of RIXS reveals that the magnon dispersion is made up of two branches well separated in energy and gapped across the entire Brillouin zone, rather than a single dominant branch. It suggests dimerization induced by the Heisenberg exchange that couples Ir ions in adjacent planes of the bilayer [105]. A better understanding of the magnetic behavior has yet to be established. (RIXS is an X-ray spectroscopy widely used for studies of electronic structures of materials in general and, in recent years, the iridates in particular. See Appendix, Section E.)

2.3.3 Transport Properties

An insulating state is illustrated in the electrical resistivity $\rho(T)$ over the range $1.7 < T < 1000$ K, as shown in **Fig. 2.22**. The insulating behavior persists up to 1000 K in $Sr_3Ir_2O_7$ [5]. However, ρ is at least four orders of magnitude smaller than that of Sr_2IrO_4. Both ρ_a and ρ_c increase slowly with temperature decreasing from 1000 to 300 K, but then rise rapidly in the vicinity of T_N and T_D, demonstrating a coupling between magnetic and transport properties [5]. The transport behavior of $Sr_3Ir_2O_7$ contrasts with that of Sr_2IrO_4, where such a correlation is absent. The differing behavior may be due in part to the difference in the charge gap Δ between $Sr_3Ir_2O_7$ ($\Delta \sim 0.18$ eV) and Sr_2IrO_4 ($\Delta \sim 0.62$ eV) and, thus, the relative effect of the AFM excitation energy on the charge gap. A Raman study also suggests different influence of frustrating exchange interactions on

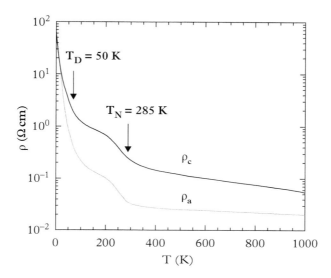

Fig. 2.22 *Sr_3Ir_2O_7: Temperature dependence of the a-axis resistivity ρ_a and the c-axis resistivity ρ_c for $1.7 < T \leq 1000$ K [5].*

Sr$_2$IrO$_4$ and Sr$_3$Ir$_2$O$_7$ [60]. Finally, like those of Sr$_2$IrO$_4$, transport properties of Sr$_3$Ir$_2$O$_7$ can be tuned electrically [106].

2.3.4 Effects of Chemical Substitution

The effects of chemical doping in Sr$_3$Ir$_2$O$_7$ are similar to those in Sr$_2$IrO$_4$. For example, 5% La doping readily precipitates a metallic state reflected in the electrical resistivity ρ of single-crystal (Sr$_{1-x}$La$_x$)$_3$Ir$_2$O$_7$, as shown in **Fig. 2.23**. The *a*-axis resistivity ρ_a (the *c*-axis resistivity ρ_c) is reduced by as much as a factor of 10^6 (10^5) at low temperatures as x evolves from 0 to 0.05, (see **Figs. 2.23a** and **2.23b**). For x = 0.05, there is a sharp

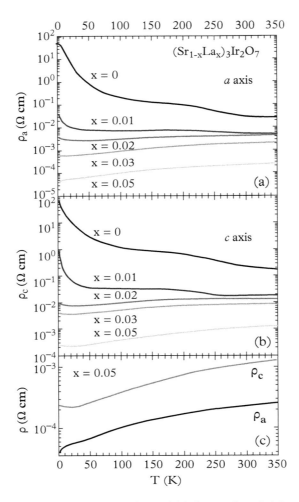

Fig. 2.23 *(Sr$_{1-x}$La$_x$)$_3$Ir$_2$O$_7$: Temperature dependence of (a) the a-axis resistivity ρ_a; and (b) the c-axis resistivity ρ_c. (c) ρ_a and ρ_c for x = 0.05 [69].*

downturn in ρ_a near 10 K, indicative of a rapid decrease in inelastic scattering (**Fig. 2.23c**). Such low-temperature behavior is also observed in oxygen-depleted $Sr_2IrO_{4-\delta}$ with $\delta = 0.04$, and La-doped Sr_2IrO_4 [10]. La doping not only adds electrons to states but also significantly increases the Ir-O-Ir bond angle θ, which is more energetically favorable for electron hopping and superexchange interactions [69].

It should be stressed that the occurrence of the metallic state in this system is not accompanied by a complete disappearance of the magnetic order, although it is significantly weakened. A study of $(Sr_{1-x}La_x)_3Ir_2O_7$ utilizing resonant elastic and inelastic X-ray scattering at the Ir-L_3 edge reveals that with increasing x, the three-dimensional long-range AFM order is gradually suppressed and evolves into a three-dimensional short-range order across the insulator-to-metal transition from x = 0 to x = 0.05, which is then followed by a transition to two-dimensional short-range order between x = 0.05 and x = 0.065. Because of the interactions between the $J_{eff} = 1/2$ pseudospins and the emergent itinerant electrons, magnetic excitations undergo damping, anisotropic softening, and gap collapse, accompanied by spin-orbit excitons that are weakly dependent on doping (**Fig. 2.24**) [107]. It is also suggested that electron doping suppresses the magnetic

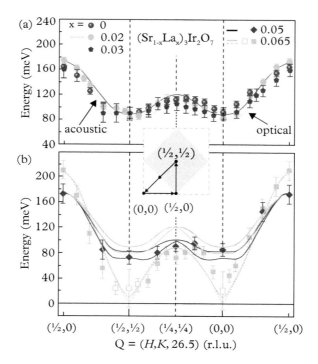

Fig. 2.24 *($Sr_{1-x}La_x)_3Ir_2O_7$: Doping-dependent magnon dispersions: (a) x = 0, 0.02, and 0.03 and (b) x = 0.05 and 0.065. The dispersions for x ≤ 0.05 and the solid squares of x = 0.065 are obtained by selecting the peak positions of the magnetic excitations. The open squares of x = 0.065 are extracted from fitting of the magnetic excitations. The solid curves are fits to the dispersions for x = 0.02, 0.05, and 0.065 using the bilayer model, which includes an acoustic and an optical branch. The dashed curve is the fitting of the dispersion for 0.065 using the J − J2 − J3 model [107].*

anisotropy and interlayer couplings and drives (Sr$_{1-x}$La$_x$)$_3$Ir$_2$O$_7$ into a correlated metallic state with two-dimensional short-range AFM order. Strong AFM fluctuations of the J$_{eff}$ = 1/2 moments persist deep in this correlated metallic state, with the magnetic gap strongly suppressed [107]. More intriguing, a subtle charge-density-wave-like Fermi surface instability is observed in the metallic region of (Sr$_{1-x}$La$_x$)$_3$Ir$_2$O$_7$ near 200 K, which shows resemblance to that observed in cuprates [108]. The absence of any signatures of a new spatial periodicity below 200 K seems to suggest an unconventional and possibly short-ranged density wave order [108].

2.3.5 Effects of High Pressure

Sr$_3$Ir$_2$O$_7$ responds to pressure in a fashion similar to that of Sr$_2$IrO$_4$ at lower pressures as discussed earlier. An early pressure study reveals an abrupt change in electrical resistivity [69] along with an apparent second-order structural change near 13 GPa [72]. A later study using both RIXS and electrical resistivity indicates that Sr$_3$Ir$_2$O$_7$ becomes a confined metal at 59.5 GPa, featuring a metallic state in the basal plane but an insulating behavior along the c axis (**Fig. 2.25**) [122]. This novel insulator-metal transition is

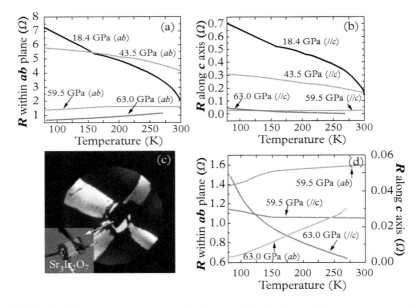

Fig. 2.25 *Sr$_3$Ir$_2$O$_7$: Electrical resistance of single-crystal Sr$_3$Ir$_2$O$_7$ at high pressure as a function of temperature for (a) the basal plane and (b) the c axis. (c) Four gold electric leads and the sample loaded into a symmetric diamond anvil cell. The inset shows two of the leads attached to the top of the single crystal, and the other two attached to the bottom. (d) Temperature dependence of the electrical resistances at 59.5 and 63.0 GPa. Note the metallic behavior in the basal plane and the insulating behavior along the c axis [122].*

attributed to a possible first-order structural change at nearby pressures [122]. The study further suggests that the structural transition above 54 GPa is likely triggered by a saturated IrO$_6$ octahedron rotation, which inevitably leads to changes in the band structure in Sr$_3$Ir$_2$O$_7$ [122]. Interestingly, a pressure-induced metallic state does not commonly occur (e.g., the persistent insulating state at 185 GPa in Sr$_2$IrO$_4$), but when it does, it has an extraordinary nature, such as in this case.

2.4 Metallic SrIrO$_3$ and Its Derivative

The semimetallic state of SrIrO$_3$ has been of great interest both theoretically and experimentally [8,43,44,109–119]. Initial work indicates that a strong SOI reduces the threshold of U for a metal-insulator transition [109,110]. A more recent theoretical study finds that an even larger critical U is required for a metal-insulator transition to occur in SrIrO$_3$, due to a combined effect of the lattice structure, strong SOI, and the protected line of Dirac nodes in the J$_{eff}$ = 1/2 bands near the Fermi level [111]. In essence, small hole and electron pockets with low densities of states that are present in SrIrO$_3$ render U less effective in driving a magnetic insulating state [111]. The rare occurrence of a semimetallic state in SrIrO$_3$ provides a unique opportunity to closely examine the intricate interplay of the SOI, U, and lattice degrees of freedom, as well as the correlation between the AFM state and the metal-insulator transition in the iridates. Indeed, tuning the relative strength of the SOI and U effectively changes the ground state in the iridates.

A large number of experimental studies of the orthorhombic perovskite SrIrO$_3$ have been conducted in recent years [8,44,109–120]. A study of epitaxy thin films of SrIrO$_3$ suggest that the iridate is an exotic narrow-band semimetal, which is attributed to a combined effect of strong SOI, dimensionality, and IrO$_6$ octahedral rotations (see **Fig. 2.26**) [43]. The partial occupation of numerous bands with strongly mixed orbital characters signals the breakdown of the single-band Mott picture that characterizes its sister compound Sr$_2$IrO$_4$, highlighting that the lattice properties can affect the relative strength between the SOI and Coulomb interaction [43,119]. Similar conclusions are also made in other film studies (e.g., [120]).

However, the bulk single-crystal SrIrO$_3$ forms only at high pressures and high temperatures [123], and little work on bulk single-crystal samples of this system has been done. Critical information concerning magnetic properties and their correlation with the electronic state, corresponding anisotropies, etc., is still lacking. A study on bulk single crystals of Ir-deficient, orthorhombic perovskite Sr$_{0.94}$Ir$_{0.77}$O$_{2.68}$ offers some insights [121]. Sr$_{0.94}$Ir$_{0.77}$O$_{2.68}$ retains the very same crystal structure as stoichiometric SrIrO$_3$, albeit with a rotation of IrO$_6$ octahedra within the basal plane by about 9.54° [121], which is a structural signature for all layered perovskite iridates [1–5,49,100].

Sr$_{0.94}$Ir$_{0.77}$O$_{2.68}$ exhibits sharp, simultaneous AFM and metal-insulator transitions at T$_N$ = 185 K with a charge gap of 0.027 eV, sharply contrasting with stoichiometric SrIrO$_3$, which is paramagnetic and semimetallic (**Fig. 2.27**). Recalling that the rotation of IrO$_6$ octahedra in Sr$_2$IrO$_4$ and Sr$_3$Ir$_2$O$_7$ [1–5] is critical in determining the AFM ground state

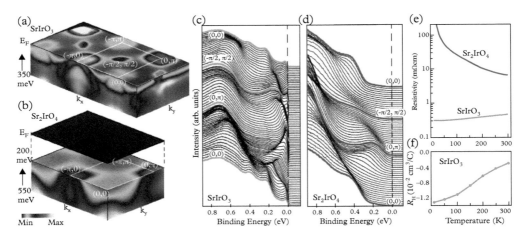

Fig. 2.26 *Epitaxial thin-film SrIrO$_3$ and Sr$_2$IrO$_4$: (a,b) E vs. k dispersions and iso-energy intensity maps of SrIrO$_3$ and Sr$_2$IrO$_4$ measured by ARPES at temperatures of 20 and 70 K, respectively. (c,d) Corresponding energy distribution curves, revealing sharp quasiparticle peaks and extremely narrow bandwidths (c) in SrIrO$_3$, compared to the broader bandwidths (d) in Sr$_2$IrO$_4$. (e) The in-plane resistivity showing the metallic and insulating transport characteristics of SrIrO$_3$ and Sr$_2$IrO$_4$, respectively. (f) The Hall resistivity of SrIrO$_3$ exhibits a large temperature dependence, consistent with the complex multiband electronic structure [43].*

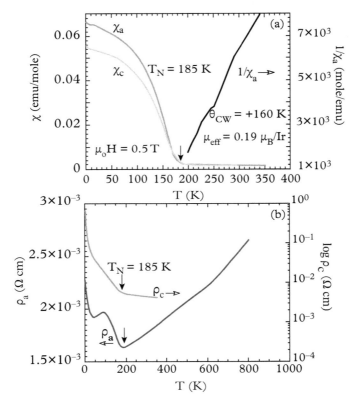

Fig. 2.27 *Sr$_{0.94}$Ir$_{0.77}$O$_{2.68}$: The temperature dependence of the magnetic susceptibility χ for a-axis χ$_a$ and c-axis χ$_c$, and χ$_a^{-1}$ (right scale) at μ$_o$H = 0.5 T, and (b) the a-axis resistivity ρ$_a$ up to 800 K, and the c-axis resistivity ρ$_c$ (right scale) [121].*

[12], one could speculate that the emerging AFM state in $Sr_{0.94}Ir_{0.77}O_{2.68}$ might be in part a result of the increased in-plane rotation of IrO_6 octahedra, $9.54°$ (compared to $8.75°$ for $SrIrO_3$ [123]). The transport properties feature an extended regime of sub-linear-temperature basal-plane resistivity between 185 K and 800 K (**Fig. 2.27b**), and an abrupt sign change in the Hall resistivity below 40 K (rather than $T_N = 185$ K) that signals a transition from hole-like to electron-like behavior with decreasing temperature [121]. These studies of $SrIrO_3$ underscore the delicacy of a metallic state that is in close proximity to an AFM insulating state, and a ground state highly sensitive to slight lattice defects. The simultaneous AFM and metal-insulator transitions illustrate a direct correlation between the AFM transition and charge gap in $SrIrO_3$, which is notably absent in Sr_2IrO_4. Nevertheless, the study on the bulk $SrIrO_3$ is in agreement with those on thin-film $SrIrO_3$ in general; that is, any subtle lattice changes can effectively vary the relative strength of the SOI and U, thus enabling changes in the ground states.

Further Reading

- Nevill Mott. *Metal-Insulator Transitions.* Taylor & Francis (1990)
- Daniel Khomskii. *Transition Metal Compounds.* Cambridge (2014)
- W. Witczak-Krempa, G. Chen, Y.B. Kim, and L. Balents. Correlated quantum phenomena in the strong spin-orbit regime. *Ann. Rev. Condens. Matter Phys* 5, 57 (2014)
- J.G. Rau, E.K.H. Lee, and H.Y. Kee. Spin-orbit physics giving rise to novel phases in correlated systems: iridates and related materials. *Ann. Rev. Condens. Matter Phys.* 7, 195 (2016)
- Gang Cao and Pedro Schlottmann. The challenge of spin-orbit-tuned ground states in iridates: a key issues review. *Reports on Progress in Physics* 81, 042502 (2018)

References

1. Huang, Q., Soubeyroux, J. L., Chmaissem, O., Sora, I. N., Santoro, A., Cava, R. J., Krajewski, J. J. & Peck, W. F. Neutron Powder Diffraction Study of the Crystal Structures of Sr_2RuO_4 and Sr_2IrO_4 at Room Temperature and at 10 K, *Journal of Solid State Chemistry* **112**, 355 (1994)
2. Cava, R. J., Batlogg, B., Kiyono, K., Takagi, H., Krajewski, J. J., Peck, W. F., Rupp, L. W. & Chen, C. H. Localized-to-itinerant electron transition in $Sr_2Ir_{1-x}Ru_xO_4$, *Phys. Rev. B* **49**, 11890 (1994)
3. Crawford, M. K., Subramanian, M. A., Harlow, R. L., Fernandez-Baca, J. A., Wang, Z. R. & Johnston, D. C. Structural and magnetic studies of Sr_2IrO_4, *Phys. Rev. B* **49**, 9198 (1994)
4. Cao, G., Bolivar, J., McCall, S., Crow, J. E. & Guertin, R. P. Weak ferromagnetism, metal-to-nonmetal transition, and negative differential resistivity in single-crystal Sr_2IrO_4, *Phy. Rev. B* **57**, R11039(R) (1998)
5. Cao, G., Xin, Y., Alexander, C. S., Crow, J. E., Schlottmann, P., Crawford, M. K., Harlow, R. L. & Marshall, W. Anomalous magnetic and transport behavior in the magnetic insulator $Sr_3Ir_2O_7$, *Phys. Rev. B* **66**, 214412 (2002)
6. Cao, G., Crow, J. E., Guertin, R. P., Henning, P. F., Homes, C. C., Strongin, M., Basov, D. N. & Lochner, E. Charge density wave formation accompanying ferromagnetic ordering in quasi-one-dimensional $BaIrO_3$, *Solid State Communications* **113**, 657 (2000)

7. Kim, B. J., Jin, H., Moon, S. J., Kim, J.-Y., Park, B.-G., Leem, C. S., Yu, J., Noh, T. W., Kim, C., Oh, S.-J., Park, J.-H., Durairaj, V., Cao, G. & Rotenberg, E. Novel J_{eff}=1/2 Mott State Induced by Relativistic Spin-Orbit Coupling in Sr_2IrO_4, *Phys. Rev. Lett.* **101**, 076402 (2008)
8. Moon, S. J., Jin, H., Kim, K. W., Choi, W. S., Lee, Y. S., Yu, J., Cao, G., Sumi, A., Funakubo, H., Bernhard, C. & Noh, T. W. Dimensionality-Controlled Insulator-Metal Transition and Correlated Metallic State in 5d Transition Metal Oxides $Sr_{n+1}Ir_nO_{3n+1}$ (n=1, 2, and ∞), *Phys. Rev. Lett.* **101**, 226402 (2008)
9. Kim, B. J., Ohsumi, H., Komesu, T., Sakai, S., Morita, T., Takagi, H. & Arima, T. Phase-Sensitive Observation of a Spin-Orbital Mott State in Sr_2IrO_4, *Science* **323**, 1329 (2009)
10. Cao, G. & Schlottmann, P. The challenge of spin–orbit-tuned ground states in iridates: a key issues review, *Rep. Prog. Phys.* **81**, 042502 (2018)
11. Liu, X., Katukuri, V. M., Hozoi, L., Yin, W.-G., Dean, M. P. M., Upton, M. H., Kim, J., Casa, D., Said, A., Gog, T., Qi, T. F., Cao, G., Tsvelik, A. M., van den Brink, J. & Hill, J. P. Testing the Validity of the Strong Spin-Orbit-Coupling Limit for Octahedrally Coordinated Iridate Compounds in a Model System Sr_3CuIrO_6, *Phys. Rev. Lett.* **109**, 157401 (2012)
12. Jackeli, G. & Khaliullin, G. Mott Insulators in the Strong Spin-Orbit Coupling Limit: From Heisenberg to a Quantum Compass and Kitaev Models, *Phys. Rev. Lett.* **102**, 017205 (2009)
13. Chern, G.-W. & Perkins, N. Large-J approach to strongly coupled spin-orbital systems, *Phys. Rev. B* **80**, 180409(R) (2009)
14. Chaloupka, J., Jackeli, G. & Khaliullin, G. Kitaev-Heisenberg Model on a Honeycomb Lattice: Possible Exotic Phases in Iridium Oxides A_2IrO_3, *Phys. Rev. Lett.* **105**, 027204 (2010)
15. Shitade, A., Katsura, H., Kuneš, J., Qi, X.-L., Zhang, S.-C. & Nagaosa, N. Quantum Spin Hall Effect in a Transition Metal Oxide Na_2IrO_3, *Phys. Rev. Lett.* **102**, 256403 (2009)
16. Witczak-Krempa, W., Chen, G., Kim, Y. B. & Balents, L. Correlated Quantum Phenomena in the Strong Spin-Orbit Regime, *Annual Review of Condensed Matter Physics* **5**, 57 (2014)
17. Rau, J. G., Lee, E. K.-H. & Kee, H.-Y. Spin-Orbit Physics Giving Rise to Novel Phases in Correlated Systems: Iridates and Related Materials, *Annual Review of Condensed Matter Physics* **7**, 195 (2016)
18. Cao, G. & DeLong, L. *Frontiers of 4d- and 5d-transition Metal Oxides.* (World Scientific, 2013).
19. Okamoto, Y., Nohara, M., Aruga-Katori, H. & Takagi, H. Spin-Liquid State in the S=1/2 Hyperkagome Antiferromagnet $Na_4Ir_3O_8$, *Phys. Rev. Lett.* **99**, 137207 (2007)
20. Takayama, T., Yaresko, A., Matsumoto, A., Nuss, J., Ishii, K., Yoshida, M., Mizuki, J. & Takagi, H. Spin-orbit coupling induced semi-metallic state in the 1/3 hole-doped hyper-kagome $Na_3Ir_3O_8$, *Scientific Reports* **4**, 6818 (2014)
21. Wang, F. & Senthil, T. Twisted Hubbard Model for Sr_2IrO_4: Magnetism and Possible High Temperature Superconductivity, *Phys. Rev. Lett.* **106**, 136402 (2011)
22. Watanabe, H., Shirakawa, T. & Yunoki, S. Monte Carlo Study of an Unconventional Superconducting Phase in Iridium Oxide J_{eff}=1/2 Mott Insulators Induced by Carrier Doping, *Phys. Rev. Lett.* **110**, 027002 (2013)
23. Kim, J., Casa, D., Upton, M. H., Gog, T., Kim, Y.-J., Mitchell, J. F., van Veenendaal, M., Daghofer, M., van den Brink, J., Khaliullin, G. & Kim, B. J. Magnetic Excitation Spectra of Sr_2IrO_4 Probed by Resonant Inelastic X-Ray Scattering: Establishing Links to Cuprate Superconductors, *Phys. Rev. Lett.* **108**, 177003 (2012)
24. Khaliullin, G., Koshibae, W. & Maekawa, S. Low Energy Electronic States and Triplet Pairing in Layered Cobaltate, *Phys. Rev. Lett.* **93**, 176401 (2004)
25. Yang, Y., Wang, W.-S., Liu, J.-G., Chen, H., Dai, J.-H. & Wang, Q.-H. Superconductivity in doped Sr_2IrO_4: A functional renormalization group study, *Phys. Rev. B* **89**, 094518 (2014)

26. You, Y.-Z., Kimchi, I. & Vishwanath, A. Doping a spin-orbit Mott insulator: Topological superconductivity from the Kitaev-Heisenberg model and possible application to (Na_2/Li_2) IrO_3, *Phys. Rev. B* **86**, 085145 (2012)

27. Meng, Z. Y., Kim,Y. B. & Kee, H.-Y. Odd-Parity Triplet Superconducting Phase in Multiorbital Materials with a Strong Spin-Orbit Coupling: Application to Doped Sr_2IrO_4, *Phys. Rev. Lett.* **113**, 177003 (2014)

28. Wan, X., Turner, A. M., Vishwanath, A. & Savrasov, S. Y. Topological semimetal and Fermi-arc surface states in the electronic structure of pyrochlore iridates, *Phys. Rev. B* **83**, 205101 (2011)

29. Singh, Y. & Gegenwart, P. Antiferromagnetic Mott insulating state in single crystals of the honeycomb lattice material Na_2IrO_3, *Phys. Rev. B* **82**, 064412 (2010)

30. Liu, X., Berlijn, T., Yin, W.-G., Ku, W., Tsvelik, A., Kim, Y.-J., Gretarsson, H., Singh, Y., Gegenwart, P. & Hill, J. P. Long-range magnetic ordering in Na_2IrO_3, *Phys. Rev. B* **83**, 220403(R) (2011)

31. Choi, S. K., Coldea, R., Kolmogorov, A. N., Lancaster, T., Mazin, I. I., Blundell, S. J., Radaelli, P. G., Singh, Y., Gegenwart, P., Choi, K. R., Cheong, S.-W., Baker, P. J., Stock, C. & Taylor, J. Spin Waves and Revised Crystal Structure of Honeycomb Iridate Na_2IrO_3, *Phys. Rev. Lett.* **108**, 127204 (2012)

32. Ye, F., Chi, S., Cao, H., Chakoumakos, B. C., Fernandez-Baca, J. A., Custelcean, R., Qi, T. F., Korneta, O. B. & Cao, G. Direct evidence of a zigzag spin-chain structure in the honeycomb lattice: A neutron and x-ray diffraction investigation of single-crystal Na_2IrO_3, *Phys. Rev. B* **85**, 180403(R) (2012)

33. Price, C. C. & Perkins, N. B. Critical Properties of the Kitaev-Heisenberg Model, *Phys. Rev. Lett.* **109**, 187201 (2012)

34. Chaloupka, J., Jackeli, G. & Khaliullin, G. Zigzag Magnetic Order in the Iridium Oxide Na_2IrO_3, *Phys. Rev. Lett.* **110**, 097204 (2013)

35. Kim, C. H., Kim, H. S., Jeong, H., Jin, H. & Yu, J. Topological Quantum Phase Transition in $5d$ Transition Metal Oxide Na_2IrO_3, *Phys. Rev. Lett.* **108**, 106401 (2012)

36. Bhattacharjee, S., Lee, S.-S. & Kim, Y. B. Spin–orbital locking, emergent pseudo-spin and magnetic order in honeycomb lattice iridates, *New J. Phys.* **14**, 073015 (2012)

37. Mazin, I. I., Jeschke, H. O., Foyevtsova, K., Valentí, R. & Khomskii, D. I. Na_2IrO_3 as a Molecular Orbital Crystal, *Phys. Rev. Lett.* **109**, 197201 (2012)

38. Rayan Serrao, C., Liu, J., Heron, J. T., Singh-Bhalla, G., Yadav, A., Suresha, S. J., Paull, R. J., Yi, D., Chu, J.-H., Trassin, M., Vishwanath, A., Arenholz, E., Frontera, C., Železný, J., Jungwirth, T., Marti, X. & Ramesh, R. Epitaxy-distorted spin-orbit Mott insulator in Sr_2IrO_4 thin films, *Phys. Rev. B* **87**, 085121 (2013)

39. Nichols, J., Korneta, O. B., Terzic, J., De Long, L. E., Cao, G., Brill, J. W. & Seo, S. S. A. Anisotropic electronic properties of a-axis-oriented Sr_2IrO_4 epitaxial thin-films, *Appl. Phys. Lett.* **103**, 131910 (2013)

40. Lupascu, A., Clancy, J. P., Gretarsson, H., Nie, Z., Nichols, J., Terzic, J., Cao, G., Seo, S. S. A., Islam, Z., Upton, M. H., Kim, J., Casa, D., Gog, T., Said, A. H., Katukuri, V. M., Stoll, H., Hozoi, L., van den Brink, J. & Kim, Y.-J. Tuning Magnetic Coupling in Sr_2IrO_4 Thin Films with Epitaxial Strain, *Phys. Rev. Lett.* **112**, 147201 (2014)

41. Yang, B.-J. & Nagaosa, N. Emergent Topological Phenomena in Thin Films of Pyrochlore Iridates, *Phys. Rev. Lett.* **112**, 246402 (2014)

42. Uchida, M., Nie, Y. F., King, P. D. C., Kim, C. H., Fennie, C. J., Schlom, D. G. & Shen, K. M. Correlated vs. conventional insulating behavior in the $\mathcal{J}_{eff}=1/2$ vs.3/2 bands in the layered iridate Ba_2IrO_4, *Phys. Rev. B* **90**, 075142 (2014)

43. Nie, Y. F., King, P. D. C., Kim, C. H., Uchida, M., Wei, H. I., Faeth, B. D., Ruf, J. P., Ruff, J. P. C., Xie, L., Pan, X., Fennie, C. J., Schlom, D. G. & Shen, K. M. Interplay of Spin-Orbit Interactions, Dimensionality, and Octahedral Rotations in Semimetallic SrIrO₃, *Phys. Rev. Lett.* **114**, 016401 (2015)

44. Matsuno, J., Ihara, K., Yamamura, S., Wadati, H., Ishii, K., Shankar, V. V., Kee, H.-Y. & Takagi, H. Engineering a Spin-Orbital Magnetic Insulator by Tailoring Superlattices, *Phys. Rev. Lett.* **114**, 247209 (2015)

45. Liu, J., Kriegner, D., Horak, L., Puggioni, D., Rayan Serrao, C., Chen, R., Yi, D., Frontera, C., Holy, V., Vishwanath, A., Rondinelli, J. M., Marti, X. & Ramesh, R. Strain-induced nonsymmorphic symmetry breaking and removal of Dirac semimetallic nodal line in an orthoperovskite iridate, *Phys. Rev. B* **93**, 085118 (2016)

46. Wang, Q., Cao, Y., Waugh, J. A., Park, S. R., Qi, T. F., Korneta, O. B., Cao, G. & Dessau, D. S. Dimensionality-controlled Mott transition and correlation effects in single-layer and bilayer perovskite iridates, *Phys. Rev. B* **87**, 245109 (2013)

47. Fujiyama, S., Ohsumi, H., Komesu, T., Matsuno, J., Kim, B. J., Takata, M., Arima, T. & Takagi, H. Two-Dimensional Heisenberg Behavior of J_{eff}=1/2 Isospins in the Paramagnetic State of the Spin-Orbital Mott Insulator Sr₂IrO₄, *Phys. Rev. Lett.* **108**, 247212 (2012)

48. Dai, J., Calleja, E., Cao, G. & McElroy, K. Local density of states study of a spin-orbit-coupling induced Mott insulator Sr₂IrO₄, *Phys. Rev. B* **90**, 041102(R) (2014)

49. Ye, F., Chi, S., Chakoumakos, B. C., Fernandez-Baca, J. A., Qi, T. & Cao, G. Magnetic and crystal structures of Sr₂IrO₄: A neutron diffraction study, *Phys. Rev. B* **87**, 140406(R) (2013).

50. Zhao, L., Torchinsky, D. H., Chu, H., Ivanov, V., Lifshitz, R., Flint, R., Qi, T., Cao, G. & Hsieh, D. Evidence of an odd-parity hidden order in a spin–orbit coupled correlated iridate, *Nature Physics* **12**, 32 (2016)

51. Wang, C., Seinige, H., Cao, G., Zhou, J.-S., Goodenough, J. B. & Tsoi, M. Electrically tunable transport in the antiferromagnetic Mott insulator Sr₂IrO₄, *Phys. Rev. B* **92**, 115136 (2015).

52. Mott, N.F. *Metal-Insulator Transitions*. (Taylor & Francis Press, 1997)

53. Imada, M., Fujimori, A. & Tokura, Y. Metal-insulator transitions, *Rev. Mod. Phys.* **70**, 1039 (1998)

54. Haskel, D., Fabbris, G., Zhernenkov, M., Kong, P. P., Jin, C. Q., Cao, G. & van Veenendaal, M. Pressure Tuning of the Spin-Orbit Coupled Ground State in Sr₂IrO₄, *Phys. Rev. Lett.* **109**, 027204 (2012)

55. Cao, G., Terzic, J., Zhao, H. D., Zheng, H., De Long, L. E. & Riseborough, P. S. Electrical Control of Structural and Physical Properties via Strong Spin-Orbit Interactions in Sr₂IrO₄, *Phys. Rev. Lett.* **120**, 017201 (2018)

56. Chikara, S., Korneta, O., Crummett, W. P., DeLong, L. E., Schlottmann, P. & Cao, G. Giant magnetoelectric effect in the J_{eff}=1/2 Mott insulator Sr₂IrO₄, *Phys. Rev. B* **80**, 140407(R) (2009)

57. Lobanov, M. V., Greenblatt, M., Caspi, E. ad N., Jorgensen, J. D., Sheptyakov, D. V., Toby, B. H., Botez, C. E. & Stephens, P. W. Crystal and magnetic structure of the Ca₃Mn₂O₇ Ruddlesden–Popper phase: neutron and synchrotron x-ray diffraction study, *J. Phys.: Condens. Matter* **16**, 5339 (2004)

58. Franke, I., Baker, P. J., Blundell, S. J., Lancaster, T., Hayes, W., Pratt, F. L. & Cao, G. Measurement of the internal magnetic field in the correlated iridates Ca₄IrO₆, Ca₅Ir₃O₁₂, Sr₃Ir₂O₇ and Sr₂IrO₄, *Phys. Rev. B* **83**, 094416 (2011)

59. Bahr, S., Alfonsov, A., Jackeli, G., Khaliullin, G., Matsumoto, A., Takayama, T., Takagi, H., Büchner, B. & Kataev, V. Low-energy magnetic excitations in the spin-orbital Mott insulator Sr₂IrO₄, *Phys. Rev. B* **89**, 180401(R) (2014)

60. Gretarsson, H., Sung, N. H., Höppner, M., Kim, B. J., Keimer, B. & Le Tacon, M. Two-Magnon Raman Scattering and Pseudospin-Lattice Interactions in Sr₂IrO₄ and Sr₃Ir₂O₇, *Phys. Rev. Lett.* **116**, 136401 (2016)

61. Ye, F., Hoffmann, C., Tian, W., Zhao, H. & Cao, G. Pseudospin-lattice coupling and electric control of the square-lattice iridate Sr_2IrO_4, *Phys. Rev. B* **102**, 115120 (2020)
62. Kim, D. J., Xia, J. & Fisk, Z. Topological surface state in the Kondo insulator samarium hexaboride, *Nature Materials* **13**, 466 (2014)
63. Wang, C., Seinige, H., Cao, G., Zhou, J.-S., Goodenough, J. B. & Tsoi, M. Anisotropic Magnetoresistance in Antiferromagnetic Sr_2IrO_4, *Phys. Rev. X* **4**, 041034 (2014)
64. Chudnovskii, F. A., Odynets, L. L., Pergament, A. L. & Stefanovich, G. B. Electroforming and Switching in Oxides of Transition Metals: The Role of Metal–Insulator Transition in the Switching Mechanism, *Journal of Solid State Chemistry* **122**, 95 (1996)
65. Guertin, R. P., Bolivar, J., Cao, G., McCall, S. & Crow, J. E. Negative differential resistivity in $Ca_3Ru_2O_7$: Unusual transport and magnetic coupling in a near-metallic system, *Solid State Communications* **107**, 263 (1998)
66. Okimura, K., Ezreena, N., Sasakawa, Y. & Sakai, J. Electric-Field-Induced Multistep Resistance Switching in Planar VO_2/c-Al_2O_3 Structure, *Jpn. J. Appl. Phys.* **48**, 065003 (2009)
67. Steckel, F., Matsumoto, A., Takayama, T., Takagi, H., Büchner, B. & Hess, C. Pseudospin transport in the $\mathcal{J}_{eff} = 1/2$ antiferromagnet Sr_2IrO_4, *EPL* **114**, 57007 (2016)
68. Chen, C., Zhou, Y., Chen, X., Han, T., An, C., Zhou, Y., Yuan, Y., Zhang, B., Wang, S., Zhang, R., Zhang, L., Zhang, C., Yang, Z., DeLong, L. E. & Cao, G. Persistent insulating state at megabar pressures in strongly spin-orbit coupled Sr_2IrO_4, *Phys. Rev. B* **101**, 144102 (2020)
69. Li, L., Kong, P. P., Qi, T. F., Jin, C. Q., Yuan, S. J., DeLong, L. E., Schlottmann, P. & Cao, G. Tuning the $\mathcal{J}_{eff} = 1/2$ insulating state via electron doping and pressure in the double-layered iridate $Sr_3Ir_2O_7$, *Phys. Rev. B* **87**, 235127 (2013)
70. Korneta, O. B., Chikara, S., Parkin, S., DeLong, L. E., Schlottmann, P. & Cao, G. Pressure-induced insulating state in $Ba_{1-x}R_xIrO_3$ (R=Gd, Eu) single crystals, *Phys. Rev. B* **81**, 045101 (2010)
71. Laguna-Marco, M. A., Fabbris, G., Souza-Neto, N. M., Chikara, S., Schilling, J. S., Cao, G. & Haskel, D. Different response of transport and magnetic properties of $BaIrO_3$ to chemical and physical pressure, *Phys. Rev. B* **90**, 014419 (2014)
72. Zhao, Z., Wang, S., Qi, T. F., Zeng, Q., Hirai, S., Kong, P. P., Li, L., Park, C., Yuan, S. J., Jin, C. Q., Cao, G. & Mao, W. L. Pressure induced second-order structural transition in $Sr_3Ir_2O_7$, *J. Phys.: Condens. Matter* **26**, 215402 (2014)
73. Zocco, D. A., Hamlin, J. J., White, B. D., Kim, B. J., Jeffries, J. R., Weir, S. T., Vohra, Y. K., Allen, J. W. & Maple, M. B. Persistent non-metallic behavior in Sr_2IrO_4 and $Sr_3Ir_2O_7$ at high pressures, *J. Phys.: Condens. Matter* **26**, 255603 (2014)
74. Donnerer, C., Feng, Z., Vale, J. G., Andreev, S. N., Solovyev, I. V., Hunter, E. C., Hanfland, M., Perry, R. S., Rønnow, H. M., McMahon, M. I., Mazurenko, V. V. & McMorrow, D. F. Pressure dependence of the structure and electronic properties of $Sr_3Ir_2O_7$, *Phys. Rev. B* **93**, 174118 (2016)
75. Xi, X., Bo, X., Xu, X. S., Kong, P. P., Liu, Z., Hong, X. G., Jin, C. Q., Cao, G., Wan, X. & Carr, G. L. Honeycomb lattice Na_2IrO_3 at high pressures: A robust spin-orbit Mott insulator, *Phys. Rev. B* **98**, 125117 (2018)
76. Hanfland, M., Syassen, K., Christensen, N. E. & Novikov, D. L. New high-pressure phases of lithium, *Nature* **408**, 174 (2000)
77. Snow, C. S., Cooper, S. L., Cao, G., Crow, J. E., Fukazawa, H., Nakatsuji, S. & Maeno, Y. Pressure-Tuned Collapse of the Mott-Like State in $Ca_{n+1}Ru_nO_{3n+1}$ (n=1, 2): Raman Spectroscopic Studies, *Phys. Rev. Lett.* **89**, 226401 (2002)
78. Drozdov, A. P., Eremets, M. I., Troyan, I. A., Ksenofontov, V. & Shylin, S. I. Conventional superconductivity at 203 kelvin at high pressures in the sulfur hydride system, *Nature* **525**, 73 (2015)

79. Haskel, D., Fabbris, G., Kim, J. H., Veiga, L. S. I., Mardegan, J. R. L., Escanhoela, C. A., Chikara, S., Struzhkin, V., Senthil, T., Kim, B. J., Cao, G. & Kim, J.-W. Possible Quantum Paramagnetism in Compressed Sr_2IrO_4, *Phys. Rev. Lett.* **124**, 067201 (2020)

80. Korneta, O. B., Qi, T., Chikara, S., Parkin, S., De Long, L. E., Schlottmann, P. & Cao, G. Electron-doped $Sr_2IrO_{4-\delta}$ ($0 \leq \delta \leq 0.04$): Evolution of a disordered $\mathcal{J}_{eff}=1/2$ Mott insulator into an exotic metallic state, *Phys. Rev. B* **82**, 115117 (2010)

81. Cao, G., Lin, X. N., Chikara, S., Durairaj, V. & Elhami, E. High-temperature weak ferromagnetism on the verge of a metallic state: Impact of dilute Sr doping on $BaIrO_3$, *Phys. Rev. B* **69**, 174418 (2004)

82. Moon, S. J., Jin, H., Choi, W. S., Lee, J. S., Seo, S. S. A., Yu, J., Cao, G., Noh, T. W. & Lee, Y. S. Temperature dependence of the electronic structure of the $\mathcal{J}_{eff}=1/2$ Mott insulator Sr_2IrO_4 studied by optical spectroscopy, *Phys. Rev. B* **80**, 195110 (2009)

83. Hsieh, D., Mahmood, F., Torchinsky, D. H., Cao, G. & Gedik, N. Observation of a metal-to-insulator transition with both Mott-Hubbard and Slater characteristics in Sr_2IrO_4 from time-resolved photocarrier dynamics, *Phys. Rev. B* **86**, 035128 (2012)

84. Cao, Y., Wang, Q., Waugh, J. A., Reber, T. J., Li, H., Zhou, X., Parham, S., Park, S.-R., Plumb, N. C., Rotenberg, E., Bostwick, A., Denlinger, J. D., Qi, T., Hermele, M. A., Cao, G. & Dessau, D. S. Hallmarks of the Mott-metal crossover in the hole-doped pseudospin-1/2 Mott insulator Sr_2IrO_4, *Nature Communications* **7**, 11367 (2016)

85. Chen, X., Hogan, T., Walkup, D., Zhou, W., Pokharel, M., Yao, M., Tian, W., Ward, T. Z., Zhao, Y., Parshall, D., Opeil, C., Lynn, J. W., Madhavan, V. & Wilson, S. D. Influence of electron doping on the ground state of $(Sr_{1-x}La_x)_2IrO_4$, *Phys. Rev. B* **92**, 075125 (2015)

86. Zhang, L., Wang, F. & Lee, D.-H. Compass impurity model of Tb substitution in Sr_2IrO_4, *Phys. Rev. B* **94**, 161118(R) (2016)

87. Zhao, H. D., Terzic, J., Zheng, H., Ni, Y. F., Zhang, Y., Ye, F., Schlottmann, P. & Cao, G. Decoupling of magnetism and electric transport in single-crystal $(Sr_{1-x}A_x)_2IrO_4$ (A = Ca or Ba), *J. Phys.: Condens. Matter* **30**, 245801 (2018)

88. Clancy, J. P., Lupascu, A., Gretarsson, H., Islam, Z., Hu, Y. F., Casa, D., Nelson, C. S., LaMarra, S. C., Cao, G. & Kim, Y.-J. Dilute magnetism and spin-orbital percolation effects in $Sr_2Ir_{1-x}Rh_xO_4$, *Phys. Rev. B* **89**, 054409 (2014)

89. Yang, J.-A., Huang, Y.-P., Hermele, M., Qi, T., Cao, G. & Reznik, D. High-energy electronic excitations in Sr_2IrO_4 observed by Raman scattering, *Phys. Rev. B* **91**, 195140 (2015)

90. Kim, Y. K., Krupin, O., Denlinger, J. D., Bostwick, A., Rotenberg, E., Zhao, Q., Mitchell, J. F., Allen, J. W. & Kim, B. J. Fermi arcs in a doped pseudospin-1/2 Heisenberg antiferromagnet, *Science* **345**, 187 (2014)

91. Liu, X., Dean, M. P. M., Meng, Z. Y., Upton, M. H., Qi, T., Gog, T., Cao, Y., Lin, J. Q., Meyers, D., Ding, H., Cao, G. & Hill, J. P. Anisotropic softening of magnetic excitations in lightly electron-doped Sr_2IrO_4, *Phys. Rev. B* **93**, 241102(R) (2016)

92. Lindquist, A. W. & Kee, H.-Y. Odd-parity superconductivity driven by octahedra rotations in iridium oxides, *Phys. Rev. B* **100**, 054512 (2019)

93. Cao, G., Zhao, H., Hu, B., Pellatz, N., Reznik, D., Schlottmann, P. & Kimchi, I. Quest for quantum states via field-altering technology, *npj Quantum Mater.* **5**, 83 (2020)

94. Jeong, J., Sidis, Y., Louat, A., Brouet, V. & Bourges, P. Time-reversal symmetry breaking hidden order in $Sr_2(Ir,Rh)O_4$, *Nature Communications* **8**, 15119 (2017)

95. Varma, C. M. Non-Fermi-liquid states and pairing instability of a general model of copper oxide metals, *Phys. Rev. B* **55**, 14554 (1997)

96. Fauqué, B., Sidis, Y., Hinkov, V., Pailhès, S., Lin, C. T., Chaud, X. & Bourges, P. Magnetic Order in the Pseudogap Phase of High-T_C Superconductors, *Phys. Rev. Lett.* **96**, 197001 (2006)

97. Carter, J.-M., Shankar, V. V., Zeb, M. A. & Kee, H.-Y. Semimetal and Topological Insulator in Perovskite Iridates, *Phys. Rev. B* **85**, 115105 (2012)

98. Kim, J., Said, A. H., Casa, D., Upton, M. H., Gog, T., Daghofer, M., Jackeli, G., van den Brink, J., Khaliullin, G. & Kim, B. J. Large Spin-Wave Energy Gap in the Bilayer Iridate $Sr_3Ir_2O_7$: Evidence for Enhanced Dipolar Interactions Near the Mott Metal-Insulator Transition, *Phys. Rev. Lett.* **109**, 157402 (2012)

99. Liu, C., Xu, S.-Y., Alidoust, N., Chang, T.-R., Lin, H., Dhital, C., Khadka, S., Neupane, M., Belopolski, I., Landolt, G., Jeng, H.-T., Markiewicz, R. S., Dil, J. H., Bansil, A., Wilson, S. D. & Hasan, M. Z. Spin-correlated electronic state on the surface of a spin-orbit Mott system, *Phys. Rev. B* **90**, 045127 (2014)

100. Hogan, T., Bjaalie, L., Zhao, L., Belvin, C., Wang, X., Van de Walle, C. G., Hsieh, D. & Wilson, S. D. Structural investigation of the bilayer iridate $Sr_3Ir_2O_7$, *Phys. Rev. B* **93**, 134110 (2016)

101. Lovesey, S. W., Khalyavin, D. D., Manuel, P., Chapon, L. C., Cao, G. & Qi, T. F. Magnetic symmetries in neutron and resonant x-ray Bragg diffraction patterns of four iridium oxides, *J. Phys.: Condens. Matter* **24**, 496003 (2012)

102. Kim, J. W., Choi, Y., Kim, J., Mitchell, J. F., Jackeli, G., Daghofer, M., van den Brink, J., Khaliullin, G. & Kim, B. J. Dimensionality Driven Spin-Flop Transition in Layered Iridates, *Phys. Rev. Lett.* **109**, 037204 (2012)

103. Carter, J.-M. & Kee, H.-Y. Microscopic theory of magnetism in $Sr_3Ir_2O_7$, *Phys. Rev. B* **87**, 014433 (2013)

104. Clancy, J. P., Plumb, K. W., Nelson, C. S., Islam, Z., Cao, G., Qi, T. & Kim, Y.-J. Field-induced magnetic behavior of the bilayer iridate $Sr_3Ir_2O_7$, *arXiv:1207.0960 [cond-mat]* (2012). at <http://arxiv.org/abs/1207.0960>

105. Moretti Sala, M., Schnells, V., Boseggia, S., Simonelli, L., Al-Zein, A., Vale, J. G., Paolasini, L., Hunter, E. C., Perry, R. S., Prabhakaran, D., Boothroyd, A. T., Krisch, M., Monaco, G., Rønnow, H. M., McMorrow, D. F. & Mila, F. Evidence of quantum dimer excitations in $Sr_3Ir_2O_7$, *Phys. Rev. B* **92**, 024405 (2015)

106. Morgan Williamson, M., Shen, S., Wang, C., Cao, G., Zhou, J., Goodenough, J. B. & Tsoi, M. Exploring the energy landscape of resistive switching in antiferromagnetic $Sr_3Ir_2O_7$, *Phys. Rev. B* **97**, 134431 (2018)

107. Lu, X., McNally, D. E., Moretti Sala, M., Terzic, J., Upton, M. H., Casa, D., Ingold, G., Cao, G. & Schmitt, T. Doping Evolution of Magnetic Order and Magnetic Excitations in $(Sr_{1-x}La_x)_3Ir_2O_7$, *Phys. Rev. Lett.* **118**, 027202 (2017)

108. Chu, H., Zhao, L., de la Torre, A., Hogan, T., Wilson, S. D. & Hsieh, D. A charge density wave-like instability in a doped spin–orbit-assisted weak Mott insulator, *Nature Materials* **16**, 200 (2017)

109. Pesin, D. & Balents, L. Mott physics and band topology in materials with strong spin–orbit interaction, *Nature Physics* **6**, 376 (2010)

110. Watanabe, H., Shirakawa, T. & Yunoki, S. Microscopic Study of a Spin-Orbit-Induced Mott Insulator in Ir Oxides, *Phys. Rev. Lett.* **105**, 216410 (2010)

111. Zeb, M. A. & Kee, H.-Y. Interplay between spin-orbit coupling and Hubbard interaction in $SrIrO_3$ and related *Pbnm* perovskite oxides, *Phys. Rev. B* **86**, 085149 (2012)

112. Chen, Y., Lu, Y.-M. & Kee, H.-Y. Topological crystalline metal in orthorhombic perovskite iridates, *Nature Communications* **6**, 6593 (2015)

113. Zhao, J. G., Yang, L. X., Yu, Y., Li, F. Y., Yu, R. C., Fang, Z., Chen, L. C. & Jin, C. Q. High-pressure synthesis of orthorhombic $SrIrO_3$ perovskite and its positive magnetoresistance, *Journal of Applied Physics* **103**, 103706 (2008)

114. Bremholm, M., K. Yim, C., Hirai, D., Climent-Pascual, E., Xu, Q., W. Zandbergen, H., N. Ali, M. & J. Cava, R. Destabilization of the 6H-$SrIrO_3$ polymorph through partial substitution of zinc and lithium, *Journal of Materials Chemistry* **22**, 16431 (2012)
115. Qasim, I., J. Kennedy, B. & Avdeev, M. Stabilising the orthorhombic perovskite structure in $SrIrO_3$ through chemical doping. Synthesis, structure and magnetic properties of $SrIr_{1-x}Mg_xO_3$ ($0.20 \leq x \leq 0.33$), *Journal of Materials Chemistry A* **1**, 13357 (2013)
116. Qasim, I., J. Kennedy, B. & Avdeev, M. Synthesis, structures and properties of transition metal doped $SrIrO_3$, *Journal of Materials Chemistry A* **1**, 3127 (2013)
117. Blanchard, P. E. R., Reynolds, E., Kennedy, B. J., Kimpton, J. A., Avdeev, M. & Belik, A. A. Anomalous thermal expansion in orthorhombic perovskite $SrIrO_3$: Interplay between spin-orbit coupling and the crystal lattice, *Phys. Rev. B* **89**, 214106 (2014)
118. Gruenewald, J. H., Nichols, J., Terzic, J., Cao, G., Brill, J. W. & Seo, S. S. A. Compressive strain-induced metal–insulator transition in orthorhombic $SrIrO_3$ thin films, *Journal of Materials Research* **29**, 2491 (2014)
119. Nie, Y. F., Di Sante, D., Chatterjee, S., King, P. D. C., Uchida, M., Ciuchi, S., Schlom, D. G. & Shen, K. M. Formation and Observation of a Quasi-Two-Dimensional d_{xy} Electron Liquid in Epitaxially Stabilized $Sr_{2-x}La_xTiO_4$ Thin Films, *Phys. Rev. Lett.* **115**, 096405 (2015)
120. Zhang, L., Jiang, X., Xu, X. & Hong, X. Abrupt enhancement of spin–orbit scattering time in ultrathin semimetallic $SrIrO_3$ close to the metal–insulator transition, *APL Materials* **8**, 051108 (2020)
121. Zheng, H., Terzic, J., Ye, F., Wan, X. G., Wang, D., Wang, J., Wang, X., Schlottmann, P., Yuan, S. J. & Cao, G. Simultaneous metal-insulator and antiferromagnetic transitions in orthorhombic perovskite iridate $Sr_{0.94}Ir_{0.78}O_{2.68}$ single crystals, *Phys. Rev. B* **93**, 235157 (2016)
122. Ding, Y., Yang, L., Chen, C.-C., Kim, H.-S., Han, M. J., Luo, W., Feng, Z., Upton, M., Casa, D., Kim, J., Gog, T., Zeng, Z., Cao, G., Mao, H. & van Veenendaal, M. Pressure-Induced Confined Metal from the Mott Insulator $Sr_3Ir_2O_7$, *Phys. Rev. Lett.* **116**, 216402 (2016)
123. Longo, J. M., Kafalas, J. A. & Arnott, R. J. Structure and properties of the high and low pressure forms of $SrIrO_3$, *Journal of Solid State Chemistry* **3**, 174 (1971)
124. Okabe, H., Isobe, M., Takayama-Muromachi, E., Koda, A., Takeshita, S., Hiraishi, M., Miyazaki, M., Kadono, R., Miyake, Y. & Akimitsu, J. Ba_2IrO_4: A spin-orbit Mott insulating quasi-two-dimensional antiferromagnet, *Phys. Rev. B* **83**, 155118 (2011)
125. Ge, M., Qi, T. F., Korneta, O. B., De Long, D. E., Schlottmann, P., Crummett, W. P. & Cao, G. Lattice-driven magnetoresistivity and metal-insulator transition in single-layered iridates, *Phys. Rev. B* **84**, 100402(R) (2011)
126. Chen, X. & Wilson, S. D. Structural evolution and electronic properties of $(Sr_{1-x}Ca_x)_{2-y}IrO_{4+z}$ spin-orbit-assisted insulators, *Phys. Rev. B* **94**, 195115 (2016)
127. Li, Q., Cao, G., Okamoto, S., Yi, J., Lin, W., Sales, B. C., Yan, J., Arita, R., Kuneš, J., Kozhevnikov, A. V., Eguiluz, A. G., Imada, M., Gai, Z., Pan, M. & Mandrus, D. G. Atomically resolved spectroscopic study of Sr_2IrO_4: Experiment and theory, *Scientific Reports* **3**, 3073 (2013)
128. Calder, S., Kim, J. W., Cao, G.-X., Cantoni, C., May, A. F., Cao, H. B., Aczel, A. A., Matsuda, M., Choi, Y., Haskel, D., Sales, B. C., Mandrus, D., Lumsden, M. D. & Christianson, A. D. Evolution of competing magnetic order in the $\mathcal{J}_{eff}=1/2$ insulating state of $Sr_2Ir_{1-x}Ru_xO_4$, *Phys. Rev. B* **92**, 165128 (2015)
129. Yuan, S. J., Aswartham, S., Terzic, J., Zheng, H., Zhao, H. D., Schlottmann, P. & Cao, G. From $\mathcal{J}_{eff}=1/2$ insulator to p-wave superconductor in single-crystal $Sr_2Ir_{1-x}Ru_xO_4$ ($0 \leq x \leq 1$), *Phys. Rev. B* **92**, 245103 (2015)
130. Qi, T. F., Korneta, O. B., Li, L., Butrouna, K., Cao, V. S., Wan, X., Schlottmann, P., Kaul, R. K. & Cao, G. Spin-orbit tuned metal-insulator transitions in single-crystal $Sr_2Ir_{1-x}Rh_xO_4$ ($0 \leq x \leq 1$), *Phys. Rev. B* **86**, 125105 (2012)

131. Gim, Y., Sethi, A., Zhao, Q., Mitchell, J. F., Cao, G. & Cooper, S. L. Isotropic and anisotropic regimes of the field-dependent spin dynamics in Sr_2IrO_4: Raman scattering studies, *Phys. Rev. B* **93**, 024405 (2016)

132. Wang, J. C., Aswartham, S., Ye, F., Terzic, J., Zheng, H., Haskel, D., Chikara, S., Choi, Y., Schlottmann, P., Custelcean, R., Yuan, S. J. & Cao, G. Decoupling of the antiferromagnetic and insulating states in Tb-doped Sr_2IrO_4, *Phys. Rev. B* **92**, 214411 (2015)

133. Cao, G., Durairaj, V., Chikara, S., DeLong, L. E., Parkin, S. & Schlottmann, P. Non-Fermi-liquid behavior in nearly ferromagnetic $SrIrO_3$ single crystals, *Phys. Rev. B* **76**, 100402(R) (2007)

134. Cao, G., Durairaj, V., Chikara, S., Parkin, S. & Schlottmann, P. Partial antiferromagnetism in spin-chain $Sr_5Rh_4O_{12}$, $Ca_5Ir_3O_{12}$ and Ca_4IrO_6 single crystals, *Phys. Rev. B* **75**, 134402 (2007)

135. Dey, T., Mahajan, A. V., Khuntia, P., Baenitz, M., Koteswararao, B. & Chou, F. C. Spin-liquid behavior in $\mathcal{J}_{eff}$=1/2 triangular lattice compound $Ba_3IrTi_2O_9$, *Phys. Rev. B* **86**, 140405(R) (2012)

136. Cao, G. unpublished.

137. Kim, S.-J., Smith, M. D., Darriet, J. & zur Loye, H.-C. Crystal growth of new perovskite and perovskite related iridates: $Ba_3LiIr_2O_9$, $Ba_3NaIr_2O_9$, and $Ba_{3.44}K_{1.56}Ir_2O_{10}$, *Journal of Solid State Chemistry* **177**, 1493 (2004)

138. Williams, S. C., Johnson, R. D., Freund, F., Choi, S., Jesche, A., Kimchi, I., Manni, S., Bombardi, A., Manuel, P., Gegenwart, P. & Coldea, R. Incommensurate counterrotating magnetic order stabilized by Kitaev interactions in the layered honeycomb α-Li_2IrO_3, *Phys. Rev. B* **93**, 195158 (2016)

139. Biffin, A., Johnson, R. D., Choi, S., Freund, F., Manni, S., Bombardi, A., Manuel, P., Gegenwart, P. & Coldea, R. Unconventional magnetic order on the hyperhoneycomb Kitaev lattice in β-Li_2IrO_3: Full solution via magnetic resonant x-ray diffraction, *Phys. Rev. B* **90**, 205116 (2014)

140. Takayama, T., Kato, A., Dinnebier, R., Nuss, J., Kono, H., Veiga, L. S. I., Fabbris, G., Haskel, D. & Takagi, H. Hyperhoneycomb Iridate β-Li_2IrO_3 as a Platform for Kitaev Magnetism, *Phys. Rev. Lett.* **114**, 077202 (2015)

141. Biffin, A., Johnson, R. D., Kimchi, I., Morris, R., Bombardi, A., Analytis, J. G., Vishwanath, A. & Coldea, R. Noncoplanar and Counterrotating Incommensurate Magnetic Order Stabilized by Kitaev Interactions in γ-Li_2IrO_3, *Phys. Rev. Lett.* **113**, 197201 (2014)

142. Modic, K. A., Smidt, T. E., Kimchi, I., Breznay, N. P., Biffin, A., Choi, S., Johnson, R. D., Coldea, R., Watkins-Curry, P., McCandless, G. T., Chan, J. Y., Gandara, F., Islam, Z., Vishwanath, A., Shekhter, A., McDonald, R. D. & Analytis, J. G. Realization of a three-dimensional spin–anisotropic harmonic honeycomb iridate, *Nature Communications* **5**, 4203 (2014)

143. Kennedy, B. J. Oxygen Vacancies in Pyrochlore Oxides: Powder Neutron Diffraction Study of $Pb_2Ir_2O_{6.5}$ and $Bi_2Ir_2O_{7-y}$, *Journal of Solid State Chemistry* **123**, 14 (1996)

144. Qi, T. F., Korneta, O. B., Wan, X., DeLong, L. E., Schlottmann, P. & Cao, G. Strong magnetic instability in correlated metallic $Bi_2Ir_2O_7$, *J. Phys.: Condens. Matter* **24**, 345601 (2012)

145. Baker, P. J., Möller, J. S., Pratt, F. L., Hayes, W., Blundell, S. J., Lancaster, T., Qi, T. F. & Cao, G. Weak magnetic transitions in pyrochlore $Bi_2Ir_2O_7$, *Phys. Rev. B* **87**, 180409(R) (2013)

146. Canals, B. & Lacroix, C. Pyrochlore Antiferromagnet: A Three-Dimensional Quantum Spin Liquid, *Phys. Rev. Lett.* **80**, 2933 (1998)

147. Gingras, M. J. P. & McClarty, P. A. Quantum spin ice: a search for gapless quantum spin liquids in pyrochlore magnets, *Rep. Prog. Phys.* **77**, 056501 (2014)

148. Cao, G., Qi, T. F., Li, L., Terzic, J., Yuan, S. J., DeLong, L. E., Murthy, G. & Kaul, R. K. Novel Magnetism of Ir^{5+} ($5d^4$) Ions in the Double Perovskite Sr_2YIrO_6, *Phys. Rev. Lett.* **112**, 056402 (2014)

149. Phelan, B. F., Seibel, E. M., Badoe, D., Xie, W. & Cava, R. J. Influence of structural distortions on the Ir magnetism in $Ba_{2-x}Sr_xYIrO_6$ double perovskites, *Solid State Communications* **236**, 37 (2016)
150. Dey, T., Maljuk, A., Efremov, D. V., Kataeva, O., Gass, S., Blum, C. G. F., Steckel, F., Gruner, D., Ritschel, T., Wolter, A. U. B., Geck, J., Hess, C., Koepernik, K., van den Brink, J., Wurmehl, S. & Büchner, B. Ba_2YIrO_6: A cubic double perovskite material with Ir^{5+} ions, *Phys. Rev. B* **93**, 014434 (2016)
151. Cao, G., Terzic, J., Zhao, H. D. & Feng, Y. unpublished. (2016)
152. Cao, G., Terzic, J., Zhao, H. D. & Feng, Y. unpublished. (2016)
153. Cao, G., Subedi, A., Calder, S., Yan, J.-Q., Yi, J., Gai, Z., Poudel, L., Singh, D. J., Lumsden, M. D., Christianson, A. D., Sales, B. C. & Mandrus, D. Magnetism and electronic structure of La_2ZnIrO_6 and La_2MgIrO_6: Candidate $\mathcal{J}_{eff}=1/2$ Mott insulators, *Phys. Rev. B* **87**, 155136 (2013)
154. Ferreira, T., Morrison, G., Yeon, J. & zur Loye, H.-C. Design and Crystal Growth of Magnetic Double Perovskite Iridates: Ln_2MIrO_6 (Ln = La, Pr, Nd, Sm-Gd; M = Mg, Ni), *Crystal Growth & Design* **16**, 2795 (2016)
155. Bremholm, M., Dutton, S. E., Stephens, P. W. & Cava, R. J. $NaIrO_3$—A pentavalent postperovskite, *Journal of Solid State Chemistry* **184**, 601 (2011)
156. Ohgushi, K., Gotou, H., Yagi, T., Kiuchi, Y., Sakai, F. & Ueda, Y. Metal-insulator transition in $Ca_{1-x}Na_xIrO_3$ with post-perovskite structure, *Phys. Rev. B* **74**, 241104(R) (2006)
157. Bogdanov, N. A., Katukuri, V. M., Stoll, H., van den Brink, J. & Hozoi, L. Post-perovskite $CaIrO_3$: A $j=1/2$ quasi-one-dimensional antiferromagnet, *Phys. Rev. B* **85**, 235147 (2012)
158. Terzic, J., Wang, J. C., Ye, F., Song, W. H., Yuan, S. J., Aswartham, S., DeLong, L. E., Streltsov, S. V., Khomskii, D. I. & Cao, G. Coexisting charge and magnetic orders in the dimer-chain iridate $Ba_5AlIr_2O_{11}$, *Phys. Rev. B* **91**, 235147 (2015)
159. Singleton, J., Kim, J. W., Topping, C. V., Hansen, A., Mun, E.-D., Chikara, S., Lakis, I., Ghannadzadeh, S., Goddard, P., Luo, X., Oh, Y. S., Cheong, S.-W. & Zapf, V. S. Magnetic properties of Sr_3NiIrO_6 and Sr_3CoIrO_6: Magnetic hysteresis with coercive fields of up to 55 T., *Phys. Rev. B* **94**, 224408 (2016)
160. Cao, G., Zheng, H., Zhao, H., Ni, Y., Pocs, C. A., Zhang, Y., Ye, F., Hoffmann, C., Wang, X., Lee, M., Hermele, M. & Kimchi, I. Quantum liquid from strange frustration in the trimer magnet $Ba_4Ir_3O_{10}$, *npj Quantum Mater.* **5**, 26 (2020)
161. Zhao, H., Ye, F., Zheng, H., Hu, B., Ni, Y., Zhang, Y., Kimchi, I. & Cao, G. Ground state in the novel dimer iridate $Ba_{13}Ir_6O_{30}$ with $Ir^{6+}(5d^3)$ ions, *Phys. Rev. B* **100**, 064418 (2019)
162. Calder, S., Cao, G.-X., Lumsden, M. D., Kim, J. W., Gai, Z., Sales, B. C., Mandrus, D. & Christianson, A. D. Magnetic structural change of Sr_2IrO_4 upon Mn doping, *Phys. Rev. B* **86**, 220403 (2012)
163. Hu, B., Zhao, H., Zhang, Y., Schlottmann, P., Ye, F. & Cao, G. Correlation between antiferromagnetic and Mott states in spin-orbit-coupled Sr_2IrO_4: A study of $Sr_2Ir_{1-x}M_xO_4$ (M=Fe or Co), *Phys. Rev. B* **103**, 115122 (2021)
164. Murayama, H., Ishida, K., Kurihara, R., Ono, T., Sato, Y., Kasahara, Y., Watanabe, H., Yanase, Y., Cao, G., Mizukami, Y., Shibauchi, T., Matsuda, Y. & Kasahara, S. Bond Directional Anapole Order in a Spin-Orbit Coupled Mott Insulator $Sr_2Ir_{1-x}Rh_xO_4$, *Phys. Rev. X* **11**, 011021 (2021)

Chapter 3

Magnetic Frustration

3.1 Overview

Quantum spin systems can enter unusual quantum phases of matter known as quantum liquids. The first such quantum liquids are discovered in one-dimensional systems that are known as *Tomonaga-Luttinger liquids*; a second class of quantum liquids can occur in two or three dimensions and are called *quantum spin liquids* [1]. These are quantum phases with fractionalized excitations that cannot be adiabatically connected to a stack of one-dimensional (1D) systems. A long-range magnetic order is often an energetically favorable state for two-dimensional (2D) and three-dimensional (3D) materials; therefore, to realize a quantum fluctuating liquid phase in 2D and 3D materials, the competing magnetic orders must be avoided through magnetic frustration. Frustration as a mechanism for quantum liquids was first discussed by Anderson [2]. One key empirical trend is that spins often prefer to anti-align with their neighbors (**Fig. 3.1a**); thus materials with triangular lattices tend to give rise to an energetic degeneracy for anti-aligned spins or frustration (**Fig. 3.1b**). Indeed, spin-liquid candidates have been found on triangular lattices such as herbertsmithite [3].

A second kind of frustration mechanism developed by Kitaev [4] is manifested in an exactly solvable spin-liquid model on the honeycomb lattice (e.g., **Fig. 3.2**) with strong spin-orbit interactions (SOI). The search for Kitaev's spin liquid has been extensive with a great deal of attention focusing on honeycomb iridates Na_2IrO_3 and Li_2IrO_3 (including β and γ phases) [5–9]. These honeycomb lattices are magnetically ordered, which suggests that the Heisenberg interaction is still sizable. There is so far no clear-cut materials realization of a quantum spin liquid in these honeycomb lattices. It is encouraging that a recent study reveals a new kind of quantum liquid arising from an unlikely place: the magnetic insulator $Ba_4Ir_3O_{10}$ where Ir_3O_{12} trimers form an *unfrustrated* square lattice [10], which is vastly different from the honeycomb and other usual frustrated lattices, offering a new pathway to realizing quantum liquids.

In this chapter, we first discuss geometrically frustrated lattices that include honeycomb iridates and ruthenates, pyrochlore systems, and double-perovskite iridates; all of them commonly feature triangular lattices as primary building blocks of their respective structures. We then turn to the new type of quantum liquid in the unfrustrated square lattice $Ba_4Ir_3O_{10}$, its mechanism and significance.

Physics of Spin-Orbit-Coupled Oxides. Gang Cao and Lance E. DeLong, Oxford University Press (2021). © Gang Cao and Lance E. DeLong.
DOI: 10.1093/oso/9780199602025.003.0003

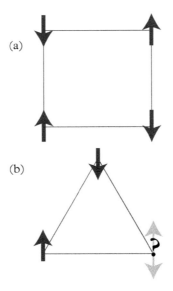

Fig. 3.1 *Schematics of antiferromagnetic exchange interaction in (a) a square lattice and (b) a triangle lattice. The question mark denotes a frustrated state.*

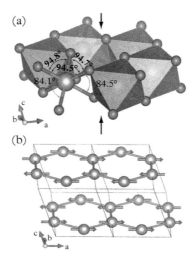

Fig. 3.2 Na_2IrO_3: *(a) Local structure within the basal plane. The compression of IrO_6 octahedron along the stacking marked by the black arrows leads to the decrease of O-Ir-O bond angles across the shared edges. (b) The zigzag order that is consistent with the symmetry associated with observed magnetic reflections. The Ir moments between honeycomb layers are antiferromagnetically coupled [14].*

3.2 Two-Dimensional Honeycomb Lattices: Na$_2$IrO$_3$ and Li$_2$IrO$_3$

The honeycomb lattices feature IrO_6 octahedra that are edge-sharing with 90° Ir-O-Ir bonds. The magnetic exchange is anisotropically bond-dependent. Such a bond-dependent interaction gives rise to strong frustration when Ir ions are placed on a honeycomb lattice and would seem to favor a Kitaev spin liquid. Theoretical treatments of the honeycomb lattices Na$_2$IrO$_3$ and Li$_2$IrO$_3$ have inspired a large body of experimental work that anticipates Kitaev physics [11–27]. If individual spins at the sites of a honeycomb lattice are restricted to align along any one of the three bond directions (six degrees of freedom for "up" and "down" spins), the Kitaev model predicts a quantum spin-liquid ground state. This novel state features short-range correlations, and the spin degrees of freedom fractionalize into Majorana fermions and a Z_2 gauge field. The honeycomb iridates are often described in terms of competing Heisenberg and Kitaev interactions; the former favors an antiferromagnetic (AFM) state, and the latter a spin-liquid state. However, no experimental confirmation of the spin-liquid state has been reported, and it is experimentally established that all known honeycomb iridates order magnetically [12–14, 21–23, 28, 29].

Na$_2$IrO$_3$ exhibits a peculiar "zigzag" magnetic order at T_N = 18 K with a Mott gap of ~0.42 eV [29], as shown in **Figs. 3.2** and **3.3**. The magnetic order was first reported in references [12,13] and later confirmed by neutron diffraction and other studies that indicate that Na$_2$IrO$_3$ orders magnetically below 18.1(2) K with Ir^{4+} ions forming zigzag spin chains within the layered honeycomb network with an ordered moment of 0.22(1)μ_B/Ir (**Fig. 3.2**) [14]. Inelastic neutron scattering on polycrystalline samples offers some insights into the magnetic state [13], whereas studies of resonant inelastic X-ray scattering (RIXS; see Appendix, Section E) characterize a branch of magnetic excitations at high-energies near 30 meV [26]. The magnetic state of Li$_2$IrO$_3$ is not as well-characterized as that of Na$_2$IrO$_3$, partly due to the lack of large single-crystal samples needed for more definitive magnetic studies. Li$_2$IrO$_3$ was initially reported to feature a paramagnetic phase [30], and then AFM order at T_N = 15 K [22]; more recently, an incommensurate magnetic order with Ir magnetic moments counter-rotating on nearest-neighbor sites was reported [31].

Indeed, the magnetic ground state of Na$_2$IrO$_3$ and Li$_2$IrO$_3$ appear not to be related. Studies of single-crystal $(Na_{1-x}Li_x)_2IrO_3$ indicate that as x is tuned, the lattice parameters evolve monotonically from Na to Li, and retain the honeycomb structure of the Ir^{4+} planes and Mott insulating state for all x. However, there is a non-monotonic, dramatic change in T_N with x in which T_N initially decreases from 18 K at x = 0 to 1.2 K at x = 0.70, before it rises to 7 K at x = 0.90 (see **Figs. 3.4** and **3.5**) [23]. Indeed, the corresponding frustration parameter $|\theta_{CW}|/T_N$ (θ_{CW} is the Curie-Weiss temperature) peaks at x = 0.70 at a value of 31.6, which suggests strongly enhanced frustration. The complicated evolution of the magnetic behavior clearly demonstrates that the magnetic ground states attained at x = 0 and 1 are indeed not related linearly [23] as had been previously suggested [17]. X-ray structure data also show that the Ir^{4+} honeycomb lattice is minimally distorted at x ≈ 0.7, where the lowest T_N and highest frustration parameter are observed.

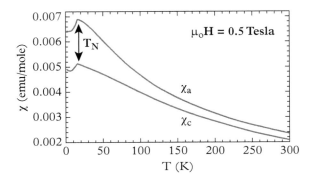

Fig. 3.3 *Na$_2$IrO$_3$: The temperature dependence of the magnetic susceptibility χ(T) for the a and c axes.*

In addition, the high-temperature anisotropy in the magnetic susceptibility is simultaneously reversed and enhanced upon Li doping. Another study of (Na$_{1-x}$Li$_x$)$_2$IrO$_3$ suggests a phase separation in the range 0.25 < x < 0.6 [32], although the specific heat data, which measure the bulk effect, indicate intrinsic changes in the magnetic state with Li doping, as shown in **Fig. 3.4**. Nevertheless, the evolution of structural, thermodynamic, and magnetic properties of (Na$_{1-x}$Li$_x$)$_2$IrO$_3$ with x suggests two possible tuning parameters for the phase transition: the crystal field splitting and the anisotropy of the distortion of the honeycomb layers, both of which change sign from the Na to Li compounds, likely compete in such a manner that explains the magnetic ordering. A relevant study of Ti-doped honeycomb lattices reveals the Curie-Weiss temperature decreases with increasing x in Na$_2$(Ir$_{1-x}$Ti$_x$)O$_3$ but remains essentially unchanged in Li$_2$(Ir$_{1-x}$Ti$_x$)O$_3$ [33], which further highlights the distinct differences between the magnetic ground states of Na$_2$IrO$_3$ and Li$_2$IrO$_3$.

There are many theoretical proposals for interactions supplementary to the Kitaev model that would cause magnetic ordering, including additional exchange processes, strong trigonal fields, and weak coupling instabilities. A consensus is yet to be reached. It has been suggested that the Kitaev spin liquid on the honeycomb lattice is extremely fragile against the second-nearest-neighbor Kitaev coupling, which is a dominant perturbation beyond the nearest-neighbor model in Na$_2$IrO$_3$ [27]. It is thought that this coupling accounts for the zigzag ordering observed in Na$_2$IrO$_3$ [27].

Chemical doping cannot induce a metallic state in the honeycomb iridates, despite extensive experimental efforts [34]. This behavior contrasts with that in the perovskite iridates, in which chemical doping can readily prompt a metallic state, as discussed in Chapter 2, and may be due to the multi-orbital nature of the honeycomb lattices.

However, like the perovskite iridates, the honeycomb iridates do not metallize at high pressures [35]. Studies suggest that this may be in part because application of pressure only broadens the bands due to the shrinking of the unit cell but does not close the energy gap (**Figs. 3.6a–3.6c**). Moreover, applied pressure causes a structural transition characterized by an abrupt change in the angle β at 3 GPa (**Figs. 3.6b** and **3.6d**). The

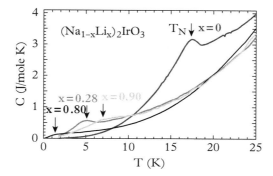

Fig. 3.4 *(Na$_{1-x}$Li$_x$)$_2$IrO$_3$: The temperature dependence of the specific heat C(T) [23].*

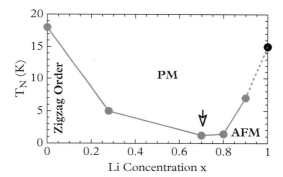

Fig. 3.5 *(Na$_{1-x}$Li$_x$)$_2$IrO$_3$: The Néel temperature T_N as a function of x. Note that the lowest T_N is 1.2 K at x = 0.7 [23].*

calculated activation energy E_g closely tracks the change of β (see **Fig. 3.6d**, right scale). This correlation between β and E_g may explain the persistent insulating state in Na₂IrO₃ [35]. Indeed, the avoidance of metallization at high pressures, which is a hallmark of the iridates studied so far, further highlights the critical role lattice distortions play in determining the ground state in the spin-orbit-coupled materials. Effects of high pressure on the layered perovskites, Sr₂IrO₄ and Sr₃Ir₂O₇, are thoroughly discussed in Chapter 2.

Na₂IrO₃ is also investigated at high magnetic fields up to 60 T using torque magnetometry measurements [36]. Remarkably, a peak-dip structure is observed in the torque response at magnetic fields corresponding to an energy scale close to T_N (see **Fig. 3.7**). A close examination suggests that such a distinctive signature in the torque is associated with dominant ferromagnetic Kitaev interactions. Furthermore, the long-range magnetic order quickly diminishes with increasing magnetic field, pointing to a possible transition to a field-induced quantum spin liquid beyond the peak-dip structure [36].

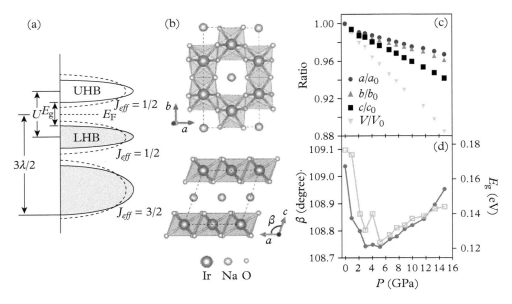

Fig. 3.6 Na_2IrO_3: *(a) A schematic illustrating the broadening (dashed lines) of the J_{eff} bands due to applied pressure. (b) The crystal structure; note the definition of the angle β. Application of pressure causes changes in the lattice parameters (c) and a pronounced change in β at 3 GPa (d); note the calculated activation gap E_g in empty squares (right scale) [35].*

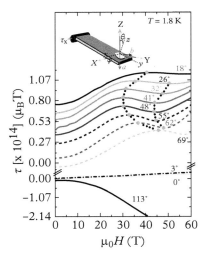

Fig. 3.7 Na_2IrO_3: *Magnetic torque, τ, as a function of magnetic field H for different polar angular orientations, θ, and azimuthal angle, $\varphi = 90°$. A peak-dip structure is observed in the magnetic torque and evolves with θ. Individual torque curves have been offset for clarity. Inset: A crystal on the cantilever with the various coordinate systems: XYZ $\rightarrow$ lab frame; xyz $\rightarrow$ frame fixed to the cantilever, so that X and x coincide. θ is the angle that the normal to the crystal makes with the magnetic field; the magnetic torque along the X direction is referred to as τ [36].*

3.3 Ruthenate Honeycomb Lattices: Na$_2$RuO$_3$ and Li$_2$RuO$_3$

The investigations of the honeycomb iridates have also spread to their ruthenate counterparts, Na$_2$RuO$_3$ and Li$_2$RuO$_3$, and more recently, Ru-based α-RuCl$_3$ (with Ru^{3+}($4d^5$)). Both of the ruthenates feature Ru^{4+}($4d^4$) ions and a weaker or "intermediate-strength" SOI (~ 0.16 eV, compared to ~ 0.4 eV for Ir ions) (see Table 1.2). The different d-shell filling and contrasting hierarchy of energy scales between the ruthenates and iridates provide a unique opportunity to gain a deeper understanding of the fundamental problem of interacting electrons on the honeycomb lattices. The magnetism of Ru^{4+} ions as well as other "heavy d^4 ions" (such as Rh^{5+}($4d^4$), Re^{3+}($5d^4$), Os^{4+}($5d^4$), and Ir^{5+}($5d^4$)) is interesting in its own right, as emphasized recently [37–43]. Materials with heavy d^4 ions tend to adopt a low-spin state because larger crystal fields often overpower the Hund's rule coupling. On the other hand, SOI with the intermediate strength may still be strong enough to impose a competing singlet ground state or a total angular momentum J$_{\text{eff}}$ = 0 state. Novel magnetic states may thus emerge when the singlet-triplet splitting (0.05–0.20 eV) becomes comparable to exchange interactions (0.05–0.10 eV) and/or non-cubic crystal fields.

Nevertheless, Na$_2$RuO$_3$ features a nearly ideal honeycomb lattice with space group *C2/m*, and orders antiferromagnetically below 30 K [43]. On the other hand, below 300 K, single-crystal Li$_2$RuO$_3$ adopts a less ideal honeycomb lattice with either *C2/m* or more distorted *P2$_1$/m* space group. Both phases exhibit a well-defined, though different magnetic state (see **Fig. 3.8**), which sharply contrasts with the singlet ground state due to dimerization observed in polycrystalline Li$_2$RuO$_3$ [44]. In general, these honeycomb lattices feature two unequal bond distances (long and short, L$_1$ and L$_s$, respectively).

The magnetic ordering temperature systematically decreases with increasing (L$_1$-L$_s$)/ L$_s$ and eventually vanishes at a critical value where dimerization emerges, leading to the singlet ground state observed in polycrystalline Li$_2$RuO$_3$. A phase diagram uncovers a direct correlation between the ground state and basal-plane distortions (lattice-tuned magnetism) in the honeycomb ruthenates, as shown in **Fig. 3.9** [43]. Although detailed magnetic structures are yet to be determined, the existing data for the honeycomb ruthenates offer some interesting observations: (1) Both Li$_2$RuO$_3$ and Li$_2$IrO$_3$ are more structurally distorted and behave with more complexities than their Na counterparts. (2) Although the SOI is expected to impose a singlet J$_{\text{eff}}$ = 0 state for Ru^{4+}($4d^4$) ions (and a J$_{\text{eff}}$ = 1/2 state for Ir^{4+}($5d^5$) ions), the observed magnetic states in the honeycomb ruthenates, as in many other ruthenates [45], indicate that the SOI is not sufficient to induce a J$_{\text{eff}}$ = 0 state. (3) It is intriguing that all honeycomb ruthenates and iridates magnetically order in a similar temperature range despite the different role of the SOI in them [45].

A Ru-based honeycomb lattice, α-RuCl$_3$, with Ru^{3+}($4d^5$) (rather than the commonplace Ru^{4+} state for ruthenates) [46,47] has drawn a great deal of attention as a candidate for the spin-liquid state [48–52]. α-RuCl$_3$, with space group *P3$_1$12*, supports a Mott state with an AFM order below T$_N$ = 7 K [53]. The magnetic structure is determined to be zigzag, similar to that of Na$_2$IrO$_3$ [53]; but α-RuCl$_3$ adopts a more ideal honeycomb lattice without the distortions found in Na$_2$IrO$_3$. This simplified structure, combined with the weaker neutron absorption cross-section of Ru compared to Ir, favors

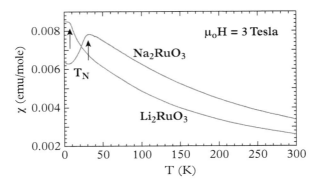

Fig. 3.8 *The temperature dependence of the magnetic susceptibility $\chi(T)$ of single-crystal Na_2RuO_3 and Li_2RuO_3.*

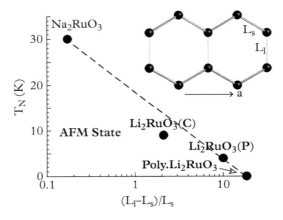

Fig. 3.9 *A phase diagram for honeycomb ruthenates: the Néel temperature T_N as a function of the bond distance ratio $(L_1-L_s)/L_s$ for all honeycomb ruthenates. Inset: A schematic of the honeycomb lattice featuring long bond L_1 and short bond L_s. Note: $C = C2/m$ and $P = P2_1/m$ [43].*

α-$RuCl_3$ for further experimental studies of Kitaev physics. Indeed, growing experimental evidence suggests that there may be a field-induced spin-liquid phase related to the physics of the Kitaev model.

3.4 Three-Dimensional Honeycomb Lattices: β-Li_2IrO_3, γ-Li_2IrO_3 and Hyperkagome $Na_4Ir_3O_8$

Two derivatives of the two-dimensional honeycomb lattices are the hyper-honeycomb Li_2IrO_3 and stripy-honeycomb Li_2IrO_3, formally termed β-Li_2IrO_3 and γ-Li_2IrO_3, respectively. They are a result of strong, mainly trigonal and monoclinic distortions of networks of edge-shared octahedra similar to those found in Na_2IrO_3 and Li_2IrO_3. With a $J_{eff} = 1/2$ state, both β-Li_2IrO_3 and γ-Li_2IrO_3 antiferromagnetically order at 38 K into incommensurate, counter-rotating spirals. But unlike the in-plane moments in Li_2IrO_3, the β- and γ-phase moments are non-coplanar, forming three-dimensional honeycomb lattices. There are some subtle differences in magnetic moments, but the two distorted honeycomb lattices are strikingly similar in terms of incommensurate propagation vectors ([0.57,0,0]) [53,54], which implies that a possible common mechanism drives the ground state. A burgeoning list of theoretical proposals emphasize a combined effect of the ferromagnetic Kitaev limit, structural distortions, and exchange interactions between nearest neighbors (or even next-nearest neighbors), which may account for the stability of the incommensurate order on the three-dimensional honeycomb lattice [27,55–59].

A large number of theoretical and experimental studies on other frustrated iridates, such as the hyperkagome $Na_4Ir_3O_8$ [60–62] and the pyrochlore iridates [6,63,64], have been conducted. $Na_4Ir_3O_8$ with a frustrated hyperkagome lattice was first reported in 2007 [60]. It features magnetic and thermal properties appropriate for a spin-liquid state (e.g., no long-range magnetic order above 2 K and no magnetic field dependence of the magnetization and heat capacity) [60]. More recent studies confirm the absence of long-range magnetic order down to T = 75 mK. This work [60] has helped generate a great deal of interest in geometrically frustrated iridates.

3.5 Pyrochlore Iridates

The pyrochlore iridates, $R_2Ir_2O_7$ (R = Y, rare earth ion or Bi or Pb), are mostly magnetic insulators [65–68]. $Bi_2Ir_2O_7$ and $Pb_2Ir_2O_7$ present a metallic ground state but are not so well studied. Many of these iridates have been intensively studied as potential platforms for exotic states such as spin liquids, Weyl semimetals, axion insulators, topological insulators and so on [6,69]. A rich phase diagram has been predicated for the pyrochlore iridates (**Fig. 3.10**) [69]. It is established that the ground state of these materials sensitively depends on the relative strengths of competing SOI and Coulomb interaction U, and a hybridization interaction that is controlled by the Ir-O-Ir bond angle [6,66,69,70]; therefore, small perturbations can easily tip the balance between the competing energies and ground states [6,66]. This is perfectly illustrated in a phase diagram presented in reference [6], in which the ground state of $R_2Ir_2O_7$ (R = rare earth ion) sensitively depends on the ionic radius of the rare earth ion [6,66]. However, the nature of the magnetic state is yet to be conclusively determined, in part due to experimental challenge of synthesizing large, high-quality single crystals of the pyrochlore iridates.

A relevant but less studied pyrochlore iridate is $Bi_2Ir_2O_7$ (where the Bi^{3+} ion substitutes for a rare earth ion) [71]. Investigations of single-crystal $Bi_2Ir_2O_7$ indicate a significantly enhanced hybridization between the Bi-$6s/6p$ and Ir-$5d$ electrons, which overpowers the SOI and U, thus driving the material into a metallic state with the

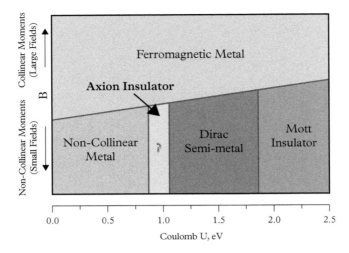

Fig. 3.10 *Pyrochlore iridates: The predicted phase diagram. The horizontal axis corresponds to U whereas the vertical axis represents external magnetic field, which can trigger a transition out of the noncollinear "all-in/all-out" ground state that includes several electronic phases [69].*

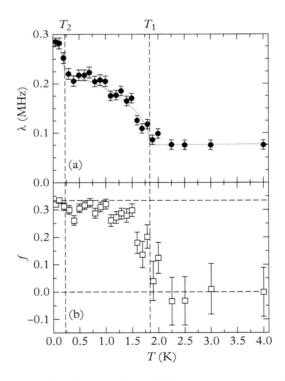

Fig. 3.11 $Bi_2Ir_2O_7$: *(a) μ-spin relaxation rate λ. The solid line is a guide to the eye. (b) Fractional amplitude of the nonrelaxing asymmetry, f. Dashed lines show transition temperatures and limiting values of f [73].*

Fermi energy residing near a sharp peak in the density of states, despite the large Z (thus SOI) for both Bi and Ir [72,73]. Muon spin relaxation (μSR; see Appendix, Section B) measurements show that $Bi_2Ir_2O_7$ undergoes a bulk magnetic transition at 1.84(3) K (**Fig. 3.11**) [73]. This is accompanied by increases in the muon spin relaxation rate and the amplitude of the nonrelaxing part of the signal, f (**Fig. 3.11b**). The magnetic field experienced by muons is estimated to be 0.7 mT at low temperatures, around two orders of magnitude smaller than that observed in other pyrochlore iridates [73]. These results suggest that the low-temperature state involves exceptionally small static magnetic moments, ~0.01μ_B/Ir. The relaxation rate increases further below 0.23(4) K, consistent with an upturn in the specific heat, suggesting the existence of a second low-temperature magnetic transition. Indeed, the coefficients (γ and β) of the low-temperature T and T^3 terms of the specific heat C(T) are strongly field dependent. The state also has a conspicuously large Wilson ratio $R_W \approx 53.5$ and an unusual Hall resistivity , ρ_H, that abruptly changes below 80 K without any correlation with the magnetic behavior (**Fig. 3.12b**) [72]. These unconventional properties, along with the novel behavior observed in metallic hexagonal $SrIrO_3$ [74], define an exotic class of SOI metals (**Fig. 3.12a**) in which strongly competing interactions induce non-Fermi liquid states that generate magnetic instabilities.

It is worth pointing out that the $5d$-electron based pyrochlore $Cd_2Re_2O_7$ is an intriguing spin-orbit-coupled material. It exhibits both a structural phase transition at 200 K

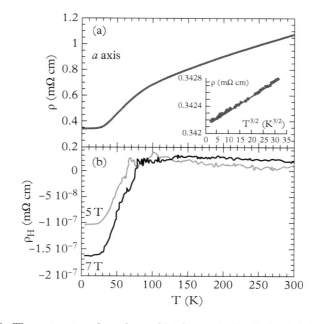

Fig. 3.12 *$Bi_2Ir_2O_7$: The temperature dependence of (a) the a-axis electrical resistivity ρ, and (b) the Hall resistivity ρ_H at 5 T and 7 T. Inset: The a-axis ρ versus $T^{3/2}$ for 1.7 K < T < 10 K [72].*

and a superconducting transition below 2 K [75,76]. Recently, a multipolar nematic phase has been discovered in $Cd_2Re_2O_7$ using spatially resolved second-harmonic optical anisotropy measurements [77]. A close examination of the critical behavior of the multipolar nematic order parameter reveals that it drives the thermal phase transition near 200 K and induces a parity-breaking lattice distortion as a secondary order [77].

3.6 Double-Perovskite Iridates with $Ir^{5+}(5d^4)$ Ions: Absence of Nonmagnetic Singlet $J_{eff} = 0$ State

The strong SOI limit is expected to lead to a nonmagnetic singlet ground state, which can be simply understood as a $J_{eff} = 0$ state arising from four electrons filling the lower $J_{eff} = 3/2$ quadruplet in materials with d^4 ions, such as $Ru^{4+}(4d^4)$, $Rh^{5+}(4d^4)$, $Re^{3+}(5d^4)$, $Os^{4+}(5d^4)$, as well as $Ir^{5+}(5d^4)$ (**Fig. 3.13c**). Indeed, the $J_{eff} = 0$ state has been used to explain the absence of magnetic ordering in the pentavalent post-perovskite $NaIrO_3$ [78], although it has also been attributed to structural distortions [40]. On the other hand, a low-spin S = 1 state is expected when U and the Hund's rule coupling J_H are much greater than the SOI λ, which is a condition commonly seen in ruthenates (**Fig. 3.13a**) [45].

Nevertheless, theoretical and experimental studies suggest that novel states in these materials can emerge when exchange interactions (0.05–0.10 eV), J_H, singlet-triplet splitting (0.050–0.20 eV), and the SOI are comparable, and therefore compete. Exotic states are expected in d^4-Mott insulators that support "intermediate-strength" SOI and U [37–39,41,42].

Pentavalent iridates attracted attention when experimental and theoretical studies showed evidence that contraindicated the anticipated $J_{eff} = 0$ state [39,41,42]. One surprising experiment addressed a distorted double-perovskite Sr_2YIrO_6 that exhibits an exotic magnetic state below 1.3 K rather than an expected $J_{eff} = 0$ state or an S = 1 state (**Fig. 3.13b**) [79]. Sr_2YIrO_6 adopts a monoclinic structure essentially derived from the $SrIrO_3$ perovskite by replacing every other Ir by nonmagnetic Y; the remaining magnetic

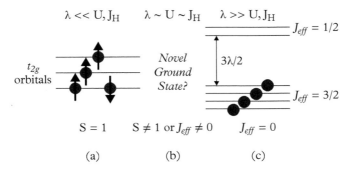

Fig. 3.13 *The ground state of $4d^4$ or $5d^4$ ions: (a) the low-spin state S = 1, and (c) the singlet $J_{eff} = 0$ state and (b) an intermediate state between S = 1 and $J_{eff} = 0$ states [84].*

Ir^{5+} ions form a network of edge-sharing tetrahedra or a face-centered cubic (FCC) structure with lattice parameters elongated compared to the parent cubic structure, as shown in **Fig. 3.14**. Because of the differences in valence state and ionic radius between Y^{3+} and Ir^{5+} ions, no significant intersite disorder is expected. This and other related double-perovskite iridates have two strongly unfavorable conditions for magnetic order, namely, pentavalent $Ir^{5+}(5d^4)$ ions, which are anticipated to have $J_{eff} = 0$ singlet ground states in the strong SOI limit, and geometric frustration in an FCC structure formed by the Ir^{5+} ions (**Fig. 3.14**).

The emergence of the unexpected magnetic ground state was initially attributed to effects of non-cubic crystal fields on the $J_{eff} = 1/2$ and $J_{eff} = 3/2$ states because such effects were not included in the original model. However, there are other possible origins for a magnetic moment in pentavalent Ir systems: as the hopping of the t_{2g} electrons increases, the width of the bands increases and the $J_{eff} = 1/2$ and $J_{eff} = 3/2$ bands may overlap so that the $J_{eff} = 1/2$ state is partially filled and the $J_{eff} = 3/2$ state has a corresponding number of holes. This may result in a magnetic moment. Interactions, in particular, the Hund's rule exchange J_H and U couple the different orbitals. Theoretical studies [37,39–42] predict a quantum phase transition with increasing hopping of the electrons from the expected $J_{eff} = 0$ state to a novel magnetic state with local $5d^4$ moments. Furthermore, a band structure study of a series of double-perovskite iridates with $Ir^{5+}(5d^4)$ ions shows

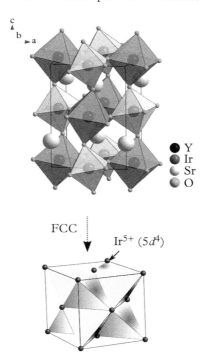

Fig. 3.14 *Sr₂YIrO₆: Upper panel: The double-perovskite crystal structure. Lower panel: The ordered replacement of nonmagnetic Y ions for magnetic Ir ions leading to a FCC lattice with geometrically frustrated edge-sharing tetrahedra formed by the pentavalent Ir⁵⁺ ions [79].*

that the e_g orbitals play no role in determining the ground state. It confirms the observed magnetic state in distorted Sr_2YIrO_6 and also predicts a breakdown of the $J_{eff} = 0$ state in undistorted, cubic Ba_2YIrO_6 as well, because the magnetic order is a result of band structure rather than of non-cubic crystal fields in these double perovskites [40]. However, there are also theoretical studies that suggest a nonmagnetic state in the double-perovskite iridates [80]; indeed, no long-range magnetic order was discerned in initial experimental studies on Ba_2YIrO_6 [81–83], although correlated magnetic moments (0.44 μ_B/Ir) were detected below 0.4 K in one of the studies, which was attributed to intermixing of Ir and Y [83].

More recent studies confirm the existence of the magnetically ordered state in the double perovskite iridates, which indicates a breakdown of the $J_{eff} = 0$ singlet state anticipated for the strong SOI limit. In fact, the unusual resilience of the magnetic order at high magnetic fields, along with the negative θ_{CW} in these materials, suggests a complex AFM configuration [79]. Moreover, similar magnetic behavior is displayed by the entire series of $(Ba_{1-x}Sr_x)_2YIrO_6$ with a similar ground state that evolves gradually with changes of the lattice parameters while retaining the underlying AFM characteristics, as shown in **Fig. 3.15** [84]. The majority of theoretical studies thus far predicts a quantum phase transition from the nonmagnetic $J_{eff} = 0$ state to a magnetic state [37,39–42]. Indeed, a more

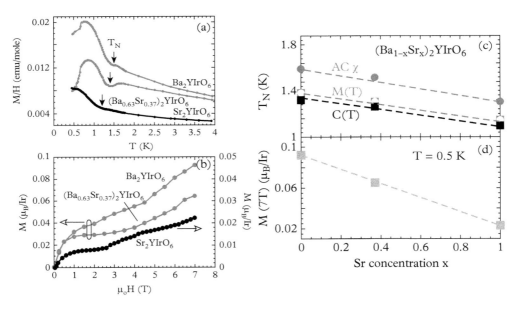

Fig. 3.15 *$(Ba_{1-x}Sr_x)_2YIrO_6$: (a) The temperature dependence of M/H. Note that the applied magnetic field $\mu_oH = 1\,T$ for $x = 0$ and $x = 1$, and $\mu_oH = 2T$ for $x = 0.37$; (b) The isothermal M(H) at $T = 0.5\,K$; (c) The magnetic transition T_N as a function of Sr doping x obtained from the data of M(T), AC $\chi(T)$, and C(T); (d) The value of the magnetization M at 7 T and 0.5 K as a function of Sr doping x [84].*

recent RIXS study reveals that the Hund's rule coupling $J_H = 0.25$ eV for Ir^{5+} in the double-perovskite iridates, suggesting that J_H should be treated on equal footing with the SOI in these materials [85]. A commonly accepted argument is that the SOI competes with a comparable Hund's rule coupling and inherently large electron hopping (because of the extended nature of $5d$-orbitals), and thus cannot stabilize the singlet ground state $J_{eff} = 0$ (**Fig. 3.13b**). However, this picture is not without controversy. A band structure study suggests that a nonmagnetic state exists in Ir^{5+}-based double perovskites with a spin gap of 200 meV [80]. A gap of this magnitude implies that such iridates should not have any significant temperature dependence in their susceptibilities at low temperatures. However, the experimental data seem to indicate otherwise.

While the $J_{eff} = 1/2$ insulating state model successfully captures the new physics observed in many iridates, recent studies suggest that it may not be adequate to describe new phenomena when the relative strength of the SOI critically competes with the strength of electron hopping and exchange interactions. It is worth mentioning once again that the $J_{eff} = 1/2$ model is a single-particle approach that assumes that Hund's rule interactions among the electrons can be neglected. Its validity needs to be closely examined when the Hund's rule interactions among the electrons are no longer negligible. Nevertheless, the anticipated magnetic ground state in the double-perovskite iridates may indicate that the SOI may not be as dominating as initially anticipated, thus leading to magnetic order rather than a singlet ground state. This magnetic state could be extraordinarily fragile as evidenced in the varied magnetic behavior reported in both experimental and theoretical studies. It is clear that the stability limits of the spin-orbit-coupled J_{eff} states in heavy transition metal materials must be investigated thoroughly.

3.7 Quantum Liquid in Unfrustrated Ba₄Ir₃O₁₀

The search for quantum spin liquids has overwhelmingly and justifiably focused on materials with triangle lattices such as honeycomb lattices (Sections 3.1 and 3.2). While the search is still going on both experimentally and theoretically, there has been so far no clear-cut materials realization of quantum spin liquids at ambient conditions.

Recently, a new, extraordinarily frustrated spin state is observed in an unlikely place— $Ba_4Ir_3O_{10}$ single crystals, where Ir_3O_{12} trimers form an unfrustrated 2D square lattice [10]. This discovery may provide an entirely new paradigm for exploring quantum liquids in spin-orbit-coupled materials with trimers.

$Ba_4Ir_3O_{10}$ adopts a monoclinic structure with a $P2_1/c$ space group [86]. The crystal structure consists of Ir_3O_{12} trimers of face-sharing IrO_6 octahedra that are vertex-linked to other trimers, forming wavelike 2D sheets in the bc-plane that are stacked along the a axis with no connectivity between the sheets (**Figs. 3.16a** and **3.16b**). The nearly identical Ir-O bond distances for both Ir1 and Ir2 sites within each trimer imply the same valence state for Ir1 and Ir2 ions in the trimers. Electrical resistivity shows a clear insulating state across the entire temperature range measured up to 400 K [10].

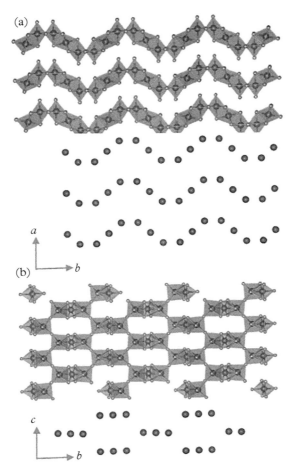

Fig. 3.16 *Crystal structure of $Ba_4Ir_3O_{10}$: (a) ab-plane. Ir sites with O octahedra at top; Ir ions highlighted below, showing the wavy 2D structure; (b) bc-plane showing the Ir trimers. Note that the resulting lattice of trimers has no geometric frustration [10].*

The magnetic susceptibility, χ, of $Ba_4Ir_3O_{10}$ for all three principal crystalline axes exhibits no anomalies or sign of magnetic order down to 1.7 K (**Fig. 3.17a**). No hysteresis behavior is seen in χ. The paramagnetic behavior perfectly follows the Curie-Weiss law for a temperature range of 100 K–350 K, which is illustrated in a plot of $1/\Delta\chi$ (**Fig. 3.17b**). Here $\Delta\chi = \chi - \chi_o$, with χ_o, the temperature-independent susceptibility, expected to arise as the Van Vleck susceptibility from high-energy crystal field levels of Ir^{4+} (with energies of order eV, much higher than the highest measurement temperatures; note that the crystal field for $5d$-electrons ranges from 2 to 5 eV; see Section 1.4). The Curie-Weiss temperature θ_{CW} (intercept on the horizontal axis) is determined to be

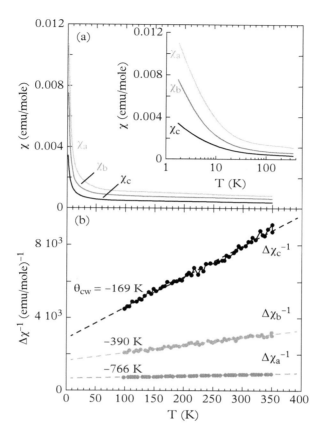

Fig. 3.17 *Magnetic susceptibility of Ba$_4$Ir$_3$O$_{10}$: Temperature dependence of (a) magnetic susceptibility χ at $\mu_o H = 1\,T$ for the a, b, and c axes. Inset: Same data in semi-log plot. (b) $1/\Delta\chi$, where $\Delta\chi = \chi - \Delta\chi_o$ with χ_o the temperature-independent (van Vleck) susceptibility, for the a, b, and c axes for $100 < T < 350\,K$. The Curie-Weiss temperatures θ_{CW} are $-766\,K$, $-390\,K$, and $-169\,K$, indicating a strong AFM inter- action and magnetic anisotropy, yet no sign of long-range order [10].*

-766 K, -390 K, and -169 K for the *a*, *b*, and *c* axes, respectively. These extracted val- ues (which vary due to the anisotropic exchange interactions) are comparable to the temperatures measured, implying that the susceptibility is not in its high-temperature asymptotic form, hence θ_{CW} should be considered as giving an order of magnitude for the exchanges. The unusually large magnitudes of θ_{CW} reveal an extraordinarily strong tendency for an AFM order—but no long-range magnetic order is discerned above 0.2 K. These results reveal an exceptionally large anisotropy-averaged frustration parameter $f = 2000$ [$= (3800+1980+840)/3$], suggesting an unusually robust, large spin- liquid regime. Moreover, the effective moment, μ_{eff}, estimated from the Curie-Weiss fit is 1.78, 0.80, and 0.4 μ_B/Ir for the *a*, *b*, and *c* axis, respectively (cf. $\mu_{eff} = 1.73\ \mu_B$ for an isolated

$S = 1/2$ moment with no SOI). These values are comparable to or greater than those of magnetically ordered iridates such as 0.13 μ_B/Ir for BaIrO$_3$ [87] and 0.50 μ_B/Ir for Sr$_2$IrO$_4$ [88], supporting the expected picture of an SOI $S_{eff} = 1/2$ local moment per Ir site.

The heat capacity, C(T), at low temperatures (0.05–4 K) shows no sign of an ordering transition above T* = 0.2 K (**Fig. 3.18**). Below 0.2 K, C(T) rises abruptly, indicating magnetic order [10]. The entropy removal below 4 K is estimated to be around 0.15 J/mole K at $\mu_o H_c = 0.2$ T, which suggests that Ba$_4$Ir$_3$O$_{10}$ behaves like a Fermi liquid metal where most of the entropy removal happens near a Fermi temperature, T$_F$, and the T-linear C(T) occurs at T << T$_F$.

C(T) presents a pronounced linear temperature dependence over a one-decade temperature span between 0.2 K and 2.5 K (**Fig. 3.18**). The linear slope is γ = 17 mJ/mole K^2. The T-linear C(T) extrapolates to a constant offset at T = 0. This behavior, which is not at all expected for any conventional insulator, reveals the existence of large residual entropy despite such low temperatures, and is consistent with a quantum liquid state. Linear heat capacity is seen in some spin-1/2 spin-liquid candidates, suggesting gapless excitations [89,90]. In addition, the thermal conductivity also features a clear linear temperature dependence below 8 K [10]. All these results point to an exotic quantum liquid state, with itinerant gapless excitations, arising at low energies from the effective SOI spin-1/2 moments in Ba$_4$Ir$_3$O$_{10}$.

This new quantum liquid state is strikingly sensitive to even slight lattice alterations, as 2% Sr substitution for Ba readily lifts the enormous frustration and stabilizes an AFM transition at T$_N$ = 130 K, shown in **Fig. 3.19**. (Note that the Sr^{2+} ion is nonmagnetic and isovalent to the Ba^{2+} ion, and that the ionic radius of Sr (1.18 Å) is smaller than that of Ba (1.36 Å); the substitution only alters the lattice parameters but causes no structural

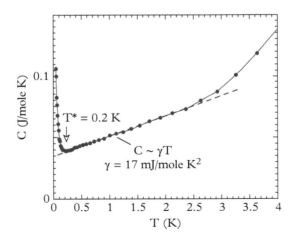

Fig. 3.18 *Low-temperature heat capacity of Ba$_4$Ir$_3$O$_{10}$: Temperature dependence of heat capacity C(T) for 0.05 K < T < 4 K. Note the linearity of C(T) [10].*

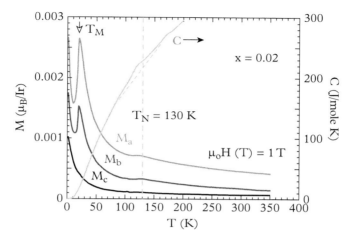

Fig. 3.19 *Magnetic properties of (Ba$_{1-x}$Sr$_x$)$_4$Ir$_3$O$_{10}$ with x = 0.02: the temperature dependence of the magnetization M for the a-, b-, and c-axes at μ$_o$H = 1 T (left scale) and heat capacity C (right scale) [10].*

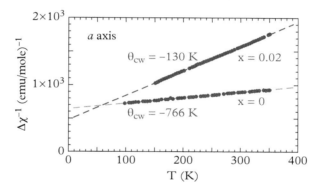

Fig. 3.20 *Magnetic properties of (Ba$_{1-x}$Sr$_x$)$_4$Ir$_3$O$_{10}$ with x = 0 and 0.02: 1/Δχ for comparison between x = 0 and x = 0.02. Note that the frustration parameter f is reduced by three orders of magnitude upon 2% Sr doping [10].*

transition.) Heat capacity C(T) is an effective measure of a bulk effect; the observed anomaly at T$_N$ = 130 K in C(T) confirms an intrinsic phase transition. C(T) shows a kink at T$_N$ but no discernable anomaly at the lower temperature T$_M$, suggesting that that peak at T$_M$ is not a sign of a phase transition. The a-axis Curie-Weiss temperature θ$_{CW}$ is changed to −130 K from −766 K, thus yielding a frustration parameter f = 1, drastically reduced from 3800 or by over three orders of magnitude, as shown in **Fig. 3.20**.

The low-temperature C(T) for x = 0.01 and 0.02 is no longer linear (**Fig. 3.21a**), with a complete removal of residual entropy below 0.05 K and the upturn below 0.2 K. The thermal conductivity along the c-axis κ$_c$ exhibits the pronounced linear dependence below 8 K for x = 0 and a distinct nonlinear feature for x = 0.02 (**Fig. 3.21b**). The behavior of κ_c is vastly different from that of other systems [91,92], and is dominated by phonons that are strongly scattered by fluctuating spins. The disappearance of the low-temperature linear dependence in $\kappa_c(T)$ (together with C(T)) upon Sr doping implies the T-linearity should be associated with the quantum liquid behavior. Moreover, $\kappa_c(T)$

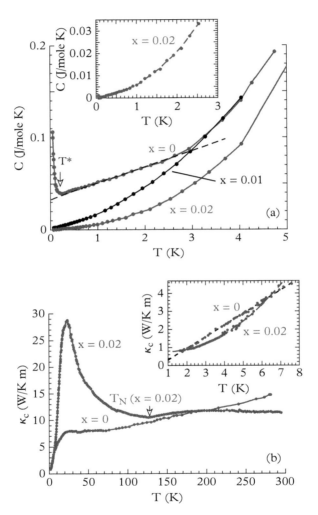

Fig. 3.21 *(Ba$_{1-x}$Sr$_x$)$_4$Ir$_3$O$_{10}$ thermal properties of x = 0.01 and 0.02 and comparison with x = 0: (a) C(T) for 0.05 K < T < 4 K for comparison between x = 0, 0.01 and 0.02. Note that the linearity that defines C(T) for x = 0 vanishes in x = 0.01 and 0.02 (inset). (b) The c-axis thermal conductivity $\kappa_c(T)$ for x = 0 and x = 0.02. Note that the drastic changes in $\kappa_c(T)$ upon Sr doping; in particular, the distinct low-temperature linearity in x = 0 and disappearance of it in x = 0.02 (inset) [10].*

for x = 0.02 shows an anomaly at T$_N$ attributable to spin-phonon scattering, followed by a peak at 25 K; the absence of such a peak in the original Ba$_4$Ir$_3$O$_{10}$ independently implies the existence of strong spin fluctuations associated with the quantum liquid state [10].

The novelty of the quantum liquid state is that frustration occurs in an unfrustrated square lattice that features Ir$_3$O$_{12}$ trimers of face-sharing IrO$_6$ octahedra. It is these trimers that form the basic magnetic unit and play a crucial role in frustration. In particular, a combined effect of the direct (Ir-Ir) and superexchange (Ir-O-Ir) interactions in the trimers results in such a delicate coupling that the middle Ir ion in a trimer is only very weakly linked to the two neighboring Ir ions (see **Fig. 3.22a**). Such "weak links" are crucial as they generate an effective one-dimensional system with zigzag chains or Luttinger liquids along the *c* axis, as illustrated in **Fig. 3.22b** [10]. Clearly, the quantum liquid state is stabilized by an extraordinarily subtle balance between the relevant

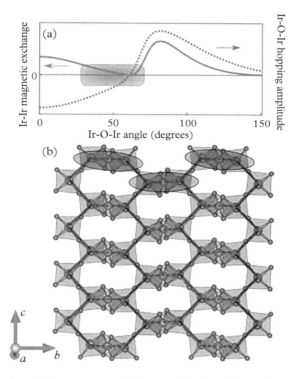

Fig. 3.22 *Hopping, trigonal distortion, and lattice recombination: (a) Ir-O-Ir hopping amplitude t for face-sharing octahedra (dotted curve) as a function of trigonally distorted Ir-O-Ir angle θ, and corresponding Ir-Ir magnetic exchange (solid curve). (θ ≈ 70.5° for ideal octahedra; θ ≈ 73° – 79° in Ba$_4$Ir$_3$O$_{10}$.) The key qualitative feature is the hopping sign change and associated region of reduced magnetic exchange (shaded oval), providing a mechanism for the lattice recombination shown in (b). (b) Trimer units (ovals on top) recombine into c-axis 1D zigzag chains (solid lines) that couple via the remaining trimer-midpoint spins (dashed lines) [10].*

interactions, and can be readily destroyed by slight chemical doping, which leads to a rapid recovery of a semiclassical magnetic order, as illustrated in **Figs. 3.19–3.21**. This discovery provides a much-needed new paradigm for the search of quantum liquids in insulators, opening the door to both new experimental avenues and to theoretical questions about newly relevant microscopic models and possible new phases of quantum matter [10].

Further Reading

- L. Savary and L. Balents. Quantum spin liquids: a review. *Rep. Prog. Phys.* 80, 016502 (2016)
- J.G. Rau, E.K.H. Lee, and H.Y. Kee. Spin-orbit physics giving rise to novel phases in correlated systems: iridates and related materials. *Ann. Rev. Condens. Matter Phys.* 7, 195 (2016)
- W. Witczak-Krempa, G. Chen, Y.B. Kim, and L. Balents. Correlated quantum phenomena in the strong spin-orbit regime. *Ann. Rev. Condens. Matter Phys.* 5, 57 (2014)

References

1. Savary, L., Balents, L. Quantum spin liquids: a review, *Rep. Prog. Phys.* 80, 016502 (2016)
2. Anderson, P.W. Resonating valence bonds: a new kind of insulator? *Mat. Res. Bull.* 8, 153 (1973)
3. Zhou, Y., Kanoda, K., Ng, T.-K. Quantum spin liquid states, *Rev. Mod. Phys.* 89, 025003 (2017)
4. Kitaev, A. Anyons in an exactly solved model and beyond, *Ann. Phys.* 321, 2 (2006)
5. Jackeli, G., Khaliullin, G. Mott insulators in the strong spin-orbit coupling limit: from Heisenberg to a quantum compass and Kitaev models, *Phys. Rev. Lett.* 102, 017205 (2009)
6. Witczak-Krempa, W., Chen, G., Kim, Y.B., Balents, L. Correlated quantum phenomena in the strong spin-orbit regime, *Ann. Rev. Condens. Matter Phys.* 5, 57 (2014)
7. Rau, J.G., Lee, E.K.H, Kee, H.-Y. Spin-orbit physics giving rise to novel phases in correlated systems: iridates and related materials, *Ann. Rev. Condens. Matter Phys.* 7, 195 (2016)
8. Cao, G., Schlottmann, P. The challenge of spin-orbit-tuned ground states in iridates: a key issues review, *Rep. Prog. Phys.* 81, 042502 (2018)
9. Hermanns, M., Kimchi, I., Knolle, J. Physics of the Kitaev model: fractionalization, dynamic correlations, and material connections, *Ann. Rev. Condens. Matter Phys.* 9, 17 (2018)
10. Cao, G., Zheng, H., Zhao, H.D., Ni, Y.F., Pocs, C.A., Zhang, Y., Ye, F., Hoffmann, C., Wang, X., Lee, M., Hermele, M., Kimchi, I. Quantum liquid from strange frustration in the trimer magnet $Ba_4Ir_3O_{10}$, *npj Quantum Materials* 5, 26 (2020)
11. Chaloupka, J., Jackeli, G., Khaliullin, G. Kitaev-Heisenberg Model on a Honeycomb Lattice: Possible Exotic Phases in Iridium Oxides A_2IrO_3, *Phys. Rev. Lett.* 105, 027204 (2010)
12. Liu, X., Berlijn, T., Yin, W.G., Ku, W., Tsvelik, A., Kim, Y.J., Gretarsson, H., Singh, Y., Gegenwart, P., Hill, J.P. Long-range magnetic ordering in Na_2IrO_3, *Phys. Rev. B* 83, 220403(R) (2011)
13. Choi, S.K., Coldea, R., Kolmogorov, A.N., Lancaster, T., Mazin, I.I., Blundell, S.J., Radaelli, P.G., Singh, Y., Gegenwart, P., Choi, K.R., Cheong, S.W., Baker, P.J., Stock, C., Taylor, J. Spin Waves and Revised Crystal Structure of Honeycomb Iridate Na_2IrO_3, *Phys. Rev. Lett.* 108, 127204 (2012)

14. Ye, F., Chi, S.X., Cao, H.B., Chakoumakos, B.C., Fernandez-Baca, J.A., Custelcean, R., Qi, T.F., Korneta, O.B., Cao, G. Direct evidence of a zigzag spin-chain structure in the honeycomb lattice: A neutron and x-ray diffraction investigation of single-crystal Na_2IrO_3, *Phys. Rev. B* 85, 180403(R) (2012)

15. Price, C.C., Perkins, N.B. Critical Properties of the Kitaev-Heisenberg Model, *Phys. Rev. Lett.* 109, 187201 (2012)

16. Chaloupka, J., Jackeli, G., Khaliullin, G. Zigzag Magnetic Order in the Iridium Oxide Na_2IrO_3, *Phys. Rev. Lett.* 110, 097204 (2013)

17. Kim, C.H., Kim, H.S., Jeong, H., Jin, H., Yu, J. Topological Quantum Phase Transition in $5d$ Transition Metal Oxide Na_2IrO_3, *Phys. Rev. Lett.* 108, 106401 (2012)

18. Bhattacharjee, S., Lee, S.S., Kim, Y.B. Spin–orbital locking, emergent pseudo-spin and magnetic order in honeycomb lattice iridates, *New J. Phys.* 14, 073015 (2012)

19. Mazin, I.I., Jeschke, H.O., Foyevtsova, K., Valenti, R., Khomskii, D.I. Na_2IrO_3 as a Molecular Orbital Crystal, *Phys. Rev. Lett.* 109, 197201 (2012)

20. Modic, K.A., Smidt, T.E., Kimchi, I., Breznay, N.P., Biffin, A., Choi, S., Johnson, R.D., Coldea, R., Watkins-Curry, P., McCandless, G.T., Chan, J.Y., Gandara, F., Islam, Z., Vishwanath, A., Shekhter, A., McDonald, R.D., Analytis, J.G. Realization of a three-dimensional spin–anisotropic harmonic honeycomb iridate, *Nat. Commun.* 5, 4203 (2014)

21. Takayama, T., Kato, A., Dinnebier, R., Nuss, J., Kono, H., I. Veiga, L.S., Fabbris, G., Haskel, D., Takagi, H. Hyperhoneycomb Iridate β-Li_2IrO_3 as a Platform for Kitaev Magnetism, *Phys. Rev. Lett.* 114, 077202 (2015)

22. Singh, Y., Manni, S., Reuther, J., Berlijn, T., Thomale, R., Ku, W., Trebst, S., Gegenwart, P. Relevance of the Heisenberg-Kitaev Model for the Honeycomb Lattice Iridates A_2IrO_3, *Phys. Rev. Lett.* 108, 127203 (2012)

23. Cao, G., Qi, T.F., Li, L., Terzic, J., Cao, V.S., Yuan, S.J., Tovar, M., Murthy, G., Kaul, R.K. Evolution of magnetism in the single-crystal honeycomb iridates $(Na_{1-x}Li_x)_2IrO_3$, *Phys. Rev. B* 88, 220414(R) (2013)

24. Chun, S.H., Kim, J.W., Kim, J., Zheng, H., Stoumpos, C.C., Malliakas, C.D., Mitchell, J.F., Mehlawat, K., Singh, Y., Choi, Y., Gog, T., Al-Zein, A., Sala, M.M., Krisch, M., Chaloupka, J., Jackeli, G., Khaliullin, G., Kim, B.J. Direct evidence for dominant bond-directional interactions in a honeycomb lattice iridate Na_2IrO_3, *Nat. Phys.* 11, 462 (2015)

25. Pratt, F.L., Baker, P.J., Blundell, S.J., Lancaster, T., Ohira-Kawamura, S., Baines, C., Shimizu, Y., Kanoda, K., Watanabe, I., Saito, G. Magnetic and non-magnetic phases of a quantum spin liquid, *Nature* 471, 612 (2011)

26. Gegenwart, P., Trebst, S. Kitaev matter, *Nat. Phys.* 11, 444 (2015)

27. Rousochatzakis, I., Reuther, J., Thomale, R., Rachel, S., Perkins, N.B. Phase Diagram and Quantum Order by Disorder in the Kitaev K_1–K_2 Honeycomb Magnet, *Phys. Rev. X* 5, 041035 (2015)

28. Gretarsson, H., Clancy, J.P., Singh, Y., Gegenwart, P., Hill, J.P., Kim, J., Upton, M.H., Said, A.H., Casa, D., Gog, T., Kim, Y.J. Magnetic excitation spectrum of Na_2IrO_3 probed with resonant inelastic x-ray scattering, *Phys. Rev. B* 87, 220407(R) (2013)

29. Comin, R., Levy, G., Ludbrook, B., Zhu, Z.H., Veenstra, C.N., Rosen, J.A., Singh, Y., Gegenwart, P., Stricker, D., Hancock, J.N., van der Marel, D., Elfimov, I.S., Damascelli, A. Na_2IrO_3 as a Novel Relativistic Mott Insulator with a 340-meV Gap, *Phys. Rev. Lett.* 109, 266406 (2012)

30. Felner, I., Bradaric, I.M. The magnetic behavior of Li_2MO_3 (M=Mn, Ru and Ir) and $Li_2(Mn_{1-x}Ru_x)O_3$, *Physica B: Condensed Matter* 311, 195 (2002)

31. Williams, S.C., Johnson, R.D., Freund, F., Choi, S., Jesche, A., Kimchi, I., Manni, S., Bombardi, A., Manuel, P., Gegenwart, P., Coldea, R. Incommensurate counterrotating magnetic order stabilized by Kitaev interactions in the layered honeycomb α-Li_2IrO_3, *Phys. Rev. B* 93, 195158 (2016)

32. Manni, S., Choi, S., Mazin, I.I., Coldea, R., Altmeyer, M., Jeschke, H.O., Valenti, R., Gegenwart, P. Effect of isoelectronic doping on the honeycomb-lattice iridate A_2IrO_3, *Phys. Rev. B* 89, 245113 (2014)

33. Manni, S., Tokiwa, Y., Gegenwart, P. Effect of nonmagnetic dilution in the honeycomb-lattice iridates Na_2IrO_3 and Li_2IrO_3, *Phys. Rev. B* 89, 241102(R) (2014)

34. Cao, G. Unpublished

35. Xi, X., Bo, X., Xu, X.S., Kong, P.P., Liu, Z., Hong, X.G., Jin, C.Q., Cao, G., Wan, X., Carr, G.L. Honeycomb lattice Na_2IrO_3 at high pressures: a robust spin-orbit Mott insulator, *Phys. Rev. B* 98, 125117 (2018)

36. Das, S.D., Kundu, S., Zhu, Z., Mun, E., McDonald, R.D., Li, G., Balicas, L., McCollam, A., Cao, G., Rau, J.G., Kee, H.Y., Tripathi, V., Sebastian, S.E. Magnetic anisotropy of the alkali iridate Na_2IrO_3 at high magnetic fields: evidence for strong ferromagnetic Kitaev correlations, *Phys. Rev. B* 99, 081101(R) (2019)

37. Khaliullin, G. Excitonic Magnetism in Van Vleck–type d^4 Mott Insulators, *Phys. Rev. Lett.* 111, 197201 (2013)

38. Chen, G., Balents, L., Schnyder, A.P. Spin-Orbital Singlet and Quantum Critical Point on the Diamond Lattice: $FeSc_2S_4$, *Phys. Rev. Lett.* 102, 096406 (2009)

39. Meetei, O.N., Cole, W.S., Randeria, M., Trivedi, N. Novel magnetic state in d^4 Mott insulators, *Phys. Rev. B* 91, 054412 (2015)

40. Bhowal, S., Baidya, S., Dasgupta, I., Saha-Dasgupta, T. Breakdown of $\mathcal{J} = 0$ nonmagnetic state in d^4 iridate double perovskites: A first-principles study, *Phys. Rev. B* 92 121113(R) (2015)

41. Kim, A.J., Jeschke, H.O., Werner, P., Valenti, R. J Freezing and Hund's Rules in Spin-Orbit-Coupled Multiorbital Hubbard Models, *Phys. Rev. Lett.* 118, 086401 (2017)

42. Svoboda, C., Randeria, M., Trivedi, N. Effective magnetic interactions in spin-orbit coupled d^4 Mott insulators, *Phys. Rev. B* 95, 014409 (2017)

43. Wang, J.C., Terzic, J., Qi, T.F., Ye, F., Yuan, S.J., Aswartham, S., Streltsov, S.V., Khomskii, D.I., Kaul, R.K., Cao, G. Lattice-tuned magnetism of $Ru^{4+}(4d^4)$ ions in single crystals of the layered honeycomb ruthenates Li_2RuO_3 and Na_2RuO_3, *Phys. Rev. B* 90, 161110(R) (2014)

44. Miura, Y., Yasui, Y., Sato, M., Igawa, N., Kakurai, K. New-Type Phase Transition of Li_2RuO_3 with Honeycomb Structure, *J. Phys. Soc. Jpn.* 76, 033705 (2007)

45. Cao, G., DeLong, L. (Eds.) *Frontiers of 4d- and 5d-transition metal oxides*. (Hackensack, NJ: World Scientific, 2013)

46. Binotto, L., Pollini, I., Spinolo, G. Optical and transport properties of the magnetic semiconductor α-$RuCl_3$, *Phys. Status Solidi B* 44, 245 (1971)

47. Kobayashi, Y., Okada, T., Asai, K., Katada, M., Sano, H., Ambe, F. Moessbauer spectroscopy and magnetization studies of α- and β-$RuCl_3$, *Inorg. Chem.* 31, 4570 (1992)

48. Sears, J.A., Songvilay, M., Plumb, K.W., Clancy, J.P., Qiu, Y., Zhao, Y., Parshall, D., Kim, Y.J. Magnetic order in α-$RuCl_3$: A honeycomb-lattice quantum magnet with strong spin-orbit coupling, *Phys. Rev. B* 91, 144420 (2015)

49. Plumb, K.W., Clancy, J.P., Sandilands, L.J., Shankar, V.V., Hu, Y.F., Burch, K.S., Kee, H.Y., Kim, Y.J. α-$RuCl_3$: A spin-orbit assisted Mott insulator on a honeycomb lattice, *Phys. Rev. B* 90, 041112(R) (2014)

50. Kubota, Y., Tanaka, H., Ono, T., Narumi, Y., Kindo, K. Successive magnetic phase transitions in α-$RuCl_3$: XY-like frustrated magnet on the honeycomb lattice, *Phys. Rev. B* 91, 094422 (2015)

51. Majumder, M., Schmidt, M., Rosner, H., Tsirlin, A.A., Yasuoka, H., Baenitz, M. Anisotropic Ru^{3+} $4d^5$ magnetism in the α-$RuCl_3$ honeycomb system: Susceptibility, specific heat, and zero-field NMR, *Phys. Rev. B* 91, 180401(R) (2015)

52. Banerjee, A., Bridges, C.A., Yan, J.Q., Aczel, A.A., Li, L., Stone, M.B., Granroth, G.E., Lumsden, M.D., Yiu, Y., Knolle, J., Bhattacharjee, S., Kovrizhin, D.L., Moessner, R., Tennant,

D.A., Mandrus, D.G., Nagler, S.E. Proximate Kitaev quantum spin liquid behaviour in a honeycomb magnet, *Nat. Mater.* 15, 733 (2016)

53. Biffin, A., Johnson, R.D., Choi, S., Freund, F., Manni, S., Bombardi, A., Manuel, P., Gegenwart, P., Coldea, R. Unconventional magnetic order on the hyperhoneycomb Kitaev lattice in β-Li$_2$IrO$_3$: Full solution via magnetic resonant x-ray diffraction, *Phys. Rev. B* 90, 205116 (2014)
54. Biffin, A., Johnson, R.D., Kimchi, I., Morris, R., Bombardi, A., Analytis, J.G., Vishwanath, A., Coldea, R. Noncoplanar and Counterrotating Incommensurate Magnetic Order Stabilized by Kitaev Interactions in γ-Li$_2$IrO$_3$, *Phys. Rev. Lett.* 113, 197201 (2014)
55. Lee, E.K.H., Schaffer, R., Bhattacharjee, S., Kim, Y.B. Heisenberg-Kitaev model on the hyperhoneycomb lattice, *Phys. Rev. B* 89, 045117 (2014)
56. Kimchi, I., Analytis, J.G., Vishwanath, A. Three-dimensional quantum spin liquids in models of harmonic-honeycomb iridates and phase diagram in an infinite-D approximation, *Phys. Rev. B* 90, 205126 (2014)
57. Lee, E.K.H., Bhattacharjee, S., Hwang, K., Kim, H.S., Jin, H., Kim, Y.B. Topological and magnetic phases with strong spin-orbit coupling on the hyperhoneycomb lattice, *Phys. Rev. B* 89, 205132 (2014)
58. Lee, S., Lee, E.K.H., Paramekanti, A., Kim, Y.B. Order-by-disorder and magnetic field response in the Heisenberg-Kitaev model on a hyperhoneycomb lattice, *Phys. Rev. B* 89, 014424 (2014)
59. Lee, E.K.H., Kim, Y.B. Theory of magnetic phase diagrams in hyperhoneycomb and harmonic-honeycomb iridates, *Phys. Rev. B* 91, 064407 (2015)
60. Okamoto, Y., Nohara, M., Aruga-Katori, H., Takagi, H. Spin-Liquid State in the S = 1/2 Hyperkagome Antiferromagnet Na$_4$Ir$_3$O$_8$, *Phys. Rev. Lett.* 99, 137207 (2007)
61. Singh, Y., Tokiwa, Y., Dong, J., Gegenwart, P. Spin liquid close to a quantum critical point in Na$_4$Ir$_3$O$_8$, *Phys. Rev. B* 88, 220413(R) (2013)
62. Podolsky, D., Kim, Y.B. Spin-orbit coupling in the metallic and spin-liquid phases of Na$_4$Ir$_3$O$_8$, *Phys. Rev. B* 83, 054401 (2011)
63. Canals, B., Lacroix, C. Pyrochlore Antiferromagnet: A Three-Dimensional Quantum Spin Liquid, *Phys. Rev. Lett.* 80, 2933 (1998)
64. Gingras, M.J.P., McClarty, P.A. Quantum spin ice: a search for gapless quantum spin liquids in pyrochlore magnets, *Rep. Prog. Phys.* 77, 056501 (2014)
65. Nakatsuji, S., Machida, Y., Maeno, Y., Tayama, T., Sakakibara, T., van Duijn, J., Balicas, L., Millican, J.N., Macaluso, R.T., Chan, J.Y. Metallic Spin-Liquid Behavior of the Geometrically Frustrated Kondo Lattice Pr$_2$Ir$_2$O$_7$, *Phys. Rev. Lett.* 96, 087204 (2006)
66. Machida, Y., Nakatsuji, S., Onoda, S., Tayama, T., Sakakibara, T. Time-reversal symmetry breaking and spontaneous Hall effect without magnetic dipole order, *Nature* 463, 210 (2010)
67. Yanagishima, D., Maeno, Y. Metal-Nonmetal Changeover in Pyrochlore Iridates, *J. Phys. Soc. Jpn.* 70, 2880 (2001)
68. Matsuhira, K., Wakeshima, M., Nakanishi, R., Yamada, T., Nakamura, A., Kawano, W., Takagi, S., Hinatsu, Y. Metal–Insulator Transition in Pyrochlore Iridates Ln$_2$Ir$_2$O$_7$ (Ln = Nd, Sm, and Eu), *J. Phys. Soc. Jpn.* 76, 043706 (2007)
69. Wan, X.G., Turner, A.M., Vishwanath, A., Savrasov, S.Y. Topological semimetal and Fermi-arc surface states in the electronic structure of pyrochlore iridates, *Phys. Rev. B* 83, 205101 (2011)
70. Yang, B.J., Kim, Y.B. Topological insulators and metal-insulator transition in the pyrochlore iridates, *Phys. Rev. B* 82, 085111 (2010)
71. Kennedy, B.J. Oxygen Vacancies in Pyrochlore Oxides: Powder Neutron Diffraction Study of Pb$_2$Ir$_2$O$_{6.5}$ and Bi$_2$Ir$_2$O$_{7-y}$, *J. Solid State Chem.* 123, 14 (1996)
72. Qi, T.F., Korneta, O.B., Wan, X.G., DeLong, L.E., Schlottmann, P., Cao, G. Strong magnetic instability in correlated metallic Bi$_2$Ir$_2$O$_7$, *J. Phys. Condens. Matter* 24, 345601 (2012)

73. Baker, P.J., Möller, J.S., Pratt, F.L., Hayes, W., Blundell, S.J., Lancaster, T., Qi, T.F., Cao, G. Weak magnetic transitions in pyrochlore $Bi_2Ir_2O_7$, *Phys. Rev. B* 87, 180409(R) (2013)

74. Cao, G., Durairaj, V., Chikara, S., DeLong, L.E., Parkin, S., Schlottmann, P. Non-Fermi-liquid behavior in nearly ferromagnetic $SrIrO_3$ single crystals, *Phys. Rev. B* 76, 100402(R) (2007)

75. Sergienko, I.A., Curnoe, S.H. Structural order parameter in the pyrochlore superconductor $Cd_2Re_2O_7$, *J. Phys. Soc. Jpn.* 72, 1607 (2003)

76. Petersen, J., Caswell, M., Dodge, J., et al. Nonlinear optical signatures of the tensor order in $Cd_2Re_2O_7$, *Nature Phys.* 2, 605 (2006)

77. Harter, J.W., Zhao, Z.Y., Yan, J.-Q., Mandrus, D.G., Hsieh, D. A parity-breaking electronic nematic phase transition in the spin-orbit coupled metal $Cd_2Re_2O_7$, *Science* 356, 295 (2017)

78. Bremholm, M., Dutton, S.E., Stephens, P.W., Cava, R.J. $NaIrO_3$—A pentavalent post-perovskite, *J. Solid State Chem.* 184, 601 (2011)

79. Cao, G., Qi, T.F., Li, L., Terzic, J., Yuan, S.J., DeLong, L.E., Murthy, G., Kaul, R.K. Novel Magnetism of Ir^{5+} ($5d^4$) Ions in the Double Perovskite Sr_2YIrO_6, *Phys. Rev. Lett.* 112, 056402 (2014)

80. Pajskr, K., Novak, P., Pokorny, V., Kolorenc, J., Arita, R., Kunes, J. On the possibility of excitonic magnetism in Ir double perovskites, *Phys. Rev. B* 93, 035129 (2016)

81. Ranjbar, B., Reynolds, E., Kayser, P., Kennedy, B.J., Hester, J.R., Kimpton, J.A. Structural and Magnetic Properties of the Iridium Double Perovskites $Ba_{2-x}Sr_xYIrO_6$, *Inorg. Chem.* 54, 10468 (2015)

82. Phelan, B.F., Seibel, E.M., Badoe, D., Xie, W.W., Cava, R.J. Influence of structural distortions on the Ir magnetism in $Ba_{2-x}Sr_xYIrO_6$ double perovskites, *Solid State Commun.* 236, 37 (2016)

83. Dey, T., Maljuk, A., Efremov, D.V., Kataeva, O., Gass, S., Blum, C.G.F., Steckel, F., Gruner, D., Ritschel, T., Wolter, A.U.B., Geck, J., Hess, C., Koepernik, K., van den Brink, J., Wurmehl, S., Buchner, B. Ba_2YIrO_6: A cubic double perovskite material with Ir^{5+} ions, *Phys. Rev. B* 93, 014434 (2016)

84. Terzic, J., Zheng, H., Ye, F., Zhao, H.D., Schlottmann, P., DeLong, L.E., Yuan, S.J., Cao, G. Evidence for a low-temperature magnetic ground state in double-perovskite iridates with $Ir^{5+}(5d^4)$ ions, *Phys. Rev. B* 96, 064436 (2017)

85. Yuan, B., Clancy, J.P., Cook, A.M., Thompson, C.M., Greedan, J., Cao, G., Jeon, B.C., Noh, T.W., Upton, M.H., Casa, D., Gog, T., Paramekanti, A., Kim, Y.J. Determination of Hund's coupling in 5d oxides using resonant inelastic x-ray scattering , *Phys. Rev. B* 95, 235114 (2017)

86. Wilkens, J., Müller-Buschbaum, H. Zur Kenntnis von $Ba_4Ir_3O_{10}$, *Z. Für Anorg. Allg. Chem.* [*J. Inorg. Gen. Chem.*] 592, 79 (1991)

87. Cao, G., Crow, J.E., Guertin, R.P., Henning, P., Homes, C.C., Strongin, M., Basov, D.N., Lochner, E. Charge density wave formation accompanying ferromagnetic ordering in quasi-one-dimensional $BaIrO_3$, *Solid State Commun.* 113, 657 (2000)

88. Cao, G., Bolivar, J., McCall, S., Crow, J.E., Guertin, R.P. Weak ferromagnetism, metal-to-nonmetal transition, and negative differential resistivity in single-crystal Sr_2IrO_4, *Phys. Rev. B* 57, R11039(R) (1998)

89. Yamashita, S., Nakazawa, Y., Oguni, M., Oshima, Y., Nojiri, H., Shimizu, Y., Miyagawa, K., Kanoda, K. Thermodynamic properties of a spin-1/2 spin-liquid state in a κ-type organic salt, *Nat. Phys.* 4, 459 (2008)

90. Anderson, P.W. The resonating valence bond state in La_2CuO_4 and superconductivity, *Science* 235, 1196 (1987)

91. Steckel, F., Matsumoto, A., Takayama, T., Takagi, H., Buchner, B., Hess, C. Pseudospin transport in the $J_{eff} = 1/2$ antiferromagnet Sr_2IrO_4, *Europhys. Lett.* 114, 57007 (2016)

92. Leahy, I.A., Pocs, C.A., Siegfried, P.E., Graf, D., Do, S.-H., Choi, K.-Y., Normand, B., Lee, M. Anomalous thermal conductivity and magnetic torque response in the honeycomb magnet α-$RuCl_3$, *Phys. Rev. Lett.* 118, 187203 (2017)

Chapter 4

Lattice-Driven Ruthenates

4.1 Overview

Ruthenates feature extended $4d$-electron orbital and comparable energy scales among on-site Coulomb interactions, crystal fields (CF), spin-orbit interactions (SOI), p-d orbital hybridization, and spin-lattice coupling. The SOI are weaker in ruthenates than in iridates but are still significant (see Table 1.2). Lattice distortions such as canting and rotations of the RuO_6 octahedra in ruthenates are particularly crucial for determining the CF level splitting and electronic band structure, and thus the system ground state (also see Section 1.4.). Physical properties of this class of materials are highly susceptible to even slight lattice distortions; as a result, external stimuli that couple to the lattice, such as magnetic field, pressure, electrical current and chemical doping, can readily generate strong or disproportional responses in structural and physical properties, leading to new electric/magnetic states or phenomena.

These characteristics are shared by many $4d$- and $5d$-transition metal materials such as iridates (discussed in Chapter 2), but they are distinctively strong and arguably most significant in ruthenates. Good examples are found in the contrasting physical properties of the Ruddlesden-Popper (PR) series $Ca_{n+1}Ru_nO_{3n+1}$ and $Sr_{n+1}Ru_nO_{3n+1}$ (n = 1, 2, 3, ∞), where n is the number of RuO layers per unit cell (see **Figs. 1.3** and **1.4**). The $Ca_{n+1}Ru_nO_{3n+1}$ compounds are all proximate to a metal-nonmetal transition and prone to antiferromagnetic (AFM) order, whereas the $Sr_{n+1}Ru_nO_{3n+1}$ compounds are metallic and tend to be ferromagnetic (FM). The stark differences primarily arise from the different ionic radii of Ca and Sr ions. Consequently, the $Ca_{n+1}Ru_nO_{3n+1}$ compounds are much more distorted than the $Sr_{n+1}Ru_nO_{3n+1}$ compounds. In addition, the Curie temperature T_C increases with n for $Sr_{n+1}Ru_nO_{3n+1}$, whereas the Néel temperature T_N decreases with n for $Ca_{n+1}Ru_nO_{3n+1}$, as shown in **Fig. 1.4** [1]. Such a drastic dependence of the ground state on the cation species has been rarely observed in other transition metal RP systems, which implies that the lattice and orbital degrees of freedom play particularly critical roles in the properties of these materials.

An examination of Table 1.4 makes it clear that ruthenates, particularly the RP series, exhibit a variety of ordered states, many of which are seldom seen in other materials. This chapter focuses on the extraordinarily high susceptibility of a few RP ruthenates to small external stimuli that readily couple to the lattice. Three exemplary topics are discussed in the following sections.

Physics of Spin-Orbit-Coupled Oxides. Gang Cao and Lance E. DeLong, Oxford University Press (2021). © Gang Cao and Lance E. DeLong.
DOI: 10.1093/oso/9780199602025.003.0004

Three exemplary topics include:

(1) Negative volume thermal expansion (NVTE) via orbital and magnetic order in doped Ca_2RuO_4 (n = 1) (*Section 4.2.*)

(2) Magneto-transport properties, including colossal magnetoresistivity via avoiding a spin-polarized state, and quantum oscillations in $Ca_3Ru_2O_7$ (n = 2) (*Section 4.3.*)

(3) Modest pressure-induced transition from ferromagnetism to antiferromagnetism in $Sr_4Ru_3O_{10}$ (n = 3) (*Section 4.4.*)

Note that Sr_2RuO_4 [2], which is arguably the most extensively studied ruthenate because of its superconductivity, is not among the compounds to be discussed in detail in this chapter. This is in part because literature on this material is widely available. Most studies on Sr_2RuO_4 focus on the superconducting parity, which was initially thought to have a *p*-wave symmetry [3]. This characterization has been challenged in recent years [4–6]. In particular, results of recent studies using nuclear magnetic resonance (NMR) spectroscopy [6] directly contradict previous NMR work that shows no change in the Knight shift below the critical temperature [7], a key piece of evidence for a *p*-wave state. In short, the nature of superconductivity in Sr_2RuO_4 is still open to debate [4–6 and references therein].

4.2 Orbital and Magnetic Order in Doped Ca_2RuO_4

Single-layered Ca_2RuO_4 (n = 1) is a member of the RP series $Ca_{n+1}Ru_nO_{3n+1}$, and is an AFM insulator with a Néel temperature T_N = 110 K [8]. It exhibits a metal-insulator transition at T_{MI} = 357 K [9], which marks a concomitant and particularly violent structural transition with a severe rotation, tilt, and flattening of RuO_6 octahedra (**Fig. 4.1a**) [9–11]. This transition effectively removes the t_{2g} orbital degeneracy (d_{xy}, d_{yz}, d_{zx}) by lowering the orbital d_{xy} relative to the other two orbitals (**Figs. 4.1b, 4.1c**); the fully populated d_{xy}-orbital, in turn, leads to an insulating state [12–20]. The *a* axis contracts by 1.5% below T_{MI}, but the *b* axis expands by 3% on cooling over a temperature interval of 250 K. The combined effect of these conflicting uniaxial thermal expansions is to drive an increasingly strong orthorhombic distortion in the basal plane, which shatters single-crystal samples and contracts the lattice volume by 1.3% as temperature is lowered from 400 K to 70 K (**Fig. 4.1a**) [9].

An abrupt AFM transition occurs only at a considerably lower Néel temperature T_N = 110 K [8], highlighting its close association with a further distorted crystal structure. Note that unlike those in many other materials, T_{MI} and T_N in both Ca_2RuO_4 and $Ca_3Ru_2O_7$ do not occur simultaneously, implying that the Coulomb interaction is not the only important driving force behind these transitions. Extensive investigations of Ca_2RuO_4 have established that the physical properties are intimately coupled to external stimuli in general, and extremely sensitive to lattice perturbations in particular [12,17–20, 22–25].

Modest Cr substitutions for Ru in Ca_2RuO_4 drastically change the metal-insulator transition T_{MI} and the magnetic behavior below T_N, and, more strikingly, it also triggers a two-step, negative volume thermal expansion (NVTE) at T_{MI} and T_N. These anomalies give rise to a total volume expansion ratio $\Delta V/V \approx 1\%$ on cooling over the interval, 90 < T < 220 K, as shown in **Fig. 4.2** [23]. Such a strong NVTE due to magnetic and orbital order is unusual; however, it is a common occurrence within a class of Mott insulators,

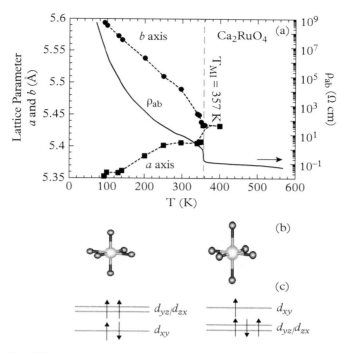

Fig. 4.1 *Ca$_2$RuO$_4$: (a) Temperature dependence of the a- and b-axis lattice parameters (left scale) and the basal-plane resistivity (right scale). Schematics illustrating (b) the difference of RuO$_6$ octahedra, and (c) the t$_{2g}$ orbitals below and above the phase transition at T$_{MI}$ = 357 K.*

Ca$_2$Ru$_{1-x}$M$_x$O$_4$, where M = Cr, Mn, Fe, and Cu, according to more extensive studies [24]. These 3d-dopants shift T$_{MI}$ by weakening the orthorhombic distortion and relaxing distortions [26], which in turn generates room for the lattice to expand, thus inducing the NVTE as well as stimulating new magnetic behavior such as metamagnetism. The observed NVTE is strongly coupled to the orbital and magnetic ordering, and sharply contrasts with the classic negative thermal expansion that is primarily driven by phonon modes or geometric effects [27–33]. In rare cases, thermal expansion is also attributed to other mechanisms such as charge transfer in BiNiO$_3$ [34]. Recently, NVTE was found in SmS, where the Kondo effect causes some of the free electrons in the metal to move into the outermost Sm electron shell to tightly screen these magnetic moments, resulting in a dramatic expansion of the material at low temperatures [35].

In the following, the NVTE observed in Cr, Mn, and Fe doped Ca$_2$RuO$_4$ is discussed along with the mechanism that drives NVTE.

4.2.1 A Novel Prototype for Negative Volume Thermal Expansion

The unusual behavior of Ca$_2$RuO$_4$ doped with Cr, Mn, and Fe provide clues concerning the mechanism that drives NVTE. A key effect of 3d-dopants on Ca$_2$RuO$_4$ is their weakening of the orthorhombic distortion. Indeed, Cr doping, while preserving the low-temperature orthorhombic symmetry (*Pbca*), reduces and eventually suppresses the orthorhombic distortion [23]. Doping is accompanied by relaxation of the Ru-O1-Ru bond angle θ, and elongation of the RuO_6 octahedra (i.e., the Ru-O2 bond distance) [23]; these structural changes facilitate correlating the NVTE with the magnetic and orbital orderings.

The close correlation of NVTE with orbital and magnetic order in $Ca_2Ru_{1-x}Cr_xO_4$ is documented by data for a representative composition x = 0.067, as shown in **Fig. 4.2**. Remarkably, strong NVTE occurs not only along the *b* axis, but also along the *a* axis for x > 0 (**Fig. 4.2a**), which presents a much rarer case of NVTE. (Note that the *a*-axis lattice parameter decreases with decreasing temperature in undoped Ca_2RuO_4, as shown in **Fig. 4.1a**.) The overall NVTE is punctuated by an abrupt expansion at T_{MI} and the

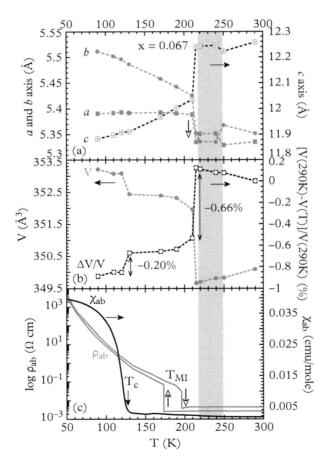

Fig. 4.2 $Ca_2Ru_{1-x}Cr_xO_4$ *with x = 0.067: The temperature dependences of: (a) lattice parameters a, b, and c axis (right scale), (b) unit cell volume V (left scale) and thermal expansion ratio $\Delta V/V$ (right scale), and (c) ab-plane resistivity ρ_{ab} and magnetic susceptibility χ_{ab} at $\mu_o H = 0.5$ T. The shaded area indicates a region of mixed tetragonal and orthorhombic phases [23].*

onset of weak FM order at T_C, and these transitions give rise to a total volume expansion $\Delta V/V \approx +1\%$ on cooling. Note that discontinuities in the *a-*, *b-*, and *c*-axis lattice parameters indicate a first-order phase transition at T_{MI} = 210 K (**Fig. 4.2a**). The unit cell volume V abruptly expands by ~ 0.66% with decreasing temperature near T_{MI}, expands again by 0.2% at T_C = 130 K, but changes only slightly between these two temperatures (**Figs. 4.2b** and **4.2c**), suggesting that an Invar effect is operative. ("Invar" refers to a specific 36%-nickel, 64%-iron alloy that has a negligible coefficient of linear thermal expansion of approximately 1.2 × 10⁻⁶ K⁻¹.)

The drastic decrease of T_{MI} observed with increasing x in $Ca_2Ru_{1-x}Cr_xO_4$, which closely tracks the rapidly weakening orthorhombic distortion, as well as the reduced tilt and elongation of RuO_6 octahedra, constitute a remarkable, correlated set of events. The disappearance of T_{MI} is concomitant with vanishing orthorhombicity for x_{cr} = 0.13 [23].

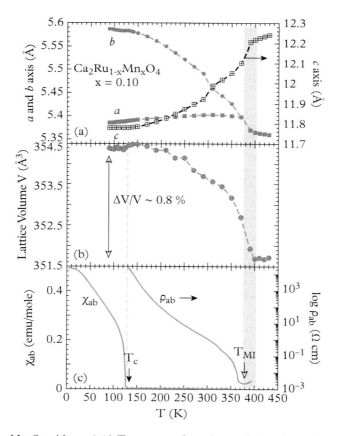

Fig. 4.3 *Ca₂Ru₁₋ₓMnₓO₄ with x = 0.10: Temperature dependences of (a) a-, b-, and c-axis lattice parameters (right scale), (b) unit cell volume V, and (c) magnetic susceptibility χ_{ab} at μ₀H = 0.5 T (field-cooled), and ab-plane resistivity log ρ_{ab} (right scale). The shaded area indicates the concomitant occurrence of the NVTE and MI transition [24].*

Moreover, the increasing Ru-O2 bond distance and Ru-O1-Ru bond angle destabilize the collinear AFM state, which, in turn, yields to weak FM behavior. A competition between AFM and FM couplings persists to x < 0.13.

The absence of NVTE in undoped Ca_2RuO_4 underscores how Cr doping softens the lattice and "unlocks" strongly buckled RuO_6 octahedra, allowing both the a and b-axis to expand while preserving the lattice symmetry. Consequently, the unit cell volume V abruptly expands on cooling just below T_{MI}, where orbital ordering occurs, and further expands at T_C. Higher Cr doping further relaxes the orthorhombic distortion that, via a highly unusual spin-lattice coupling, weakens the AFM state, and results in an extraordinary increase in volume on cooling.

Although unusual, NVTE is ubiquitous in Ca_2RuO_4 doped with other 3d-elements such as Mn [24], as shown in **Fig. 4.3a**. For x = 0.10, NVTE occurs along both the a and b axes in $Ca_2Ru_{1-x}Mn_xO_4$; this combined effect results in an overall volume expansion ratio $\Delta V/V \approx +0.8\%$ on cooling, as shown in **Fig. 4.3b**. The coupling of the NVTE to T_{MI} and T_C is obvious in that V rapidly expands below T_{MI} = 380 K and exhibits a weak but well-defined anomaly near T_C = 130 K (**Fig. 4.3c**), where the transition to weak FM order takes place, as shown in **Fig. 4.3**.

4.2.2 General Trends

We note that Fe or Cu substitutions for Ru in Ca_2RuO_4 also effectively weaken the orthorhombic distortion and induce NVTE [24], which suggests that more general conclusions can be drawn from the NVTE data. For undoped Ca_2RuO_4, the a axis decreases with decreasing temperature, especially near T_{MI} (**Fig. 4.4a**). Substituting any one of several 3d-ion types for Ru induces a modest and yet critical negative thermal expansion along the a axis near T_{MI} (**Fig. 4.4a**). We conclude that such doping is the precursor to the sizable negative thermal expansion ratio $\Delta V/V$ observed on cooling in $Ca_2Ru_{1-x}M_xO_4$. The fact that this phenomenon does not occur in *undoped* Ca_2RuO_4, despite the strong effect of the linear negative thermal expansion along the b axis, underscores how critical is the role the 3d-ion dopant M plays in unlocking the strongly buckled Ru/MO_6 octahedra, and changing the t_{2g} orbital configuration in the basal plane.

For a representative compound, $Ca_2Ru_{1-x}Fe_xO_4$ with x = 0.08, the basal plane Ru/M-O1-Ru/M bond angle θ drastically decreases below T_{MI}, which in turn prompts a simultaneous expansion of the Ru/M-O1 bond length d and the Ru/M-Ru/M distance on cooling, as shown in **Figs. 4.4b** and **4.4c**, respectively. The expansion of d clearly outweighs positive thermal expansion due to longitudinal vibrational modes, allowing both the a- and b-axis to expand with cooling, while preserving the structural symmetry. As a result, V rapidly expands on cooling near T_{MI} where the orbital order takes place.

This understanding is further strengthened by a neutron diffraction study on $Ca_2Ru_{1-x}Fe_xO_4$ [26]. This study reveals that both Ru-O(1) and Ru-O(2) bond lengths decrease as Fe doping increases (**Fig. 4.5a**); as a result, the octahedron shrinks. The shrinkage of the octahedron volume brought by Fe doping relaxes its buckled position and thus reduces the orthorhombic strain (**Figs. 4.5b–4.5d**). In particular, **Figs. 4.5c** and **4.5d**

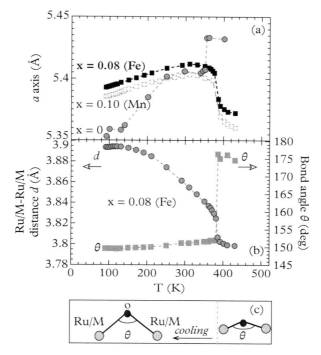

Fig. 4.4 *Temperature dependences of: (a) a-axis lattice parameter for x = 0, 0.10 (Mn), and 0.08 (Fe); (b) the Ru/M-Ru/M bond length d and the Ru/M-O1-Ru/M bond angle θ (right scale) for x = 0.08 (Fe); (c) schematics illustrating changes of d and θ on cooling [24].*

show that both the planar rotation φ and the tilt θ decrease with increasing Fe doping (the symmetry allows independent tilts of the basal O(1) plane and the apical O(2) axis). Consequently, the Ru-O-Ru bond angle increases with increasing Fe doping, as shown in **Fig. 4.5b**. In short, the 3*d*-ion dopant unlocks the severely buckled Ru/MO₆ octahedra, facilitating the NVTE.

Both the NVTE and the orbital order in $Ca_2Ru_{1-x}M_xO_4$ closely track the changing orthorhombicity as x changes, and disappear when the orthorhombicity vanishes near x_c, a critical doping concentration (e.g., x_c = 0.14, 0.25, 0.22, and 0.20 for Cr, Mn, Fe, and Cu, respectively) [23,24,26]. It is recognized that an increase in the Ru-O2 bond length along the c axis destabilizes the collinear AFM state [12], resulting in strongly competing AFM and FM exchange interactions or spin canting below T_C in doped Ca_2RuO_4.

It is also noted that the magnitude of the NVTE decreases with increasing atomic number of 3*d*-ions. This interesting trend may be associated with the fact that with increasing nuclear charge the 3*d*-orbitals become more contracted, and the 3*d*-band progressively fills and downshifts away from the Fermi energy E_F, thus weakening the overlap with the 4*d*-band that remains near E_F.

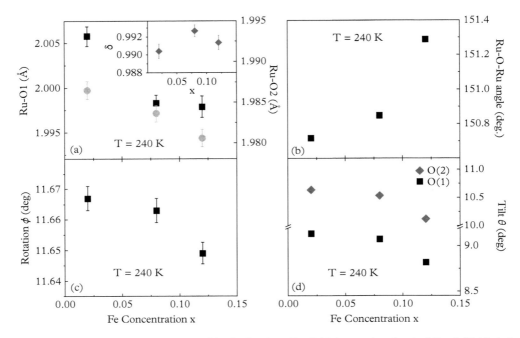

Fig. 4.5 *Fe-concentration dependence of (a) the in-plane Ru-O(1) (squares) and apical Ru-O(2) (circles) bond length, (b) the Ru-O-Ru bond angle, (c) the rotation angle of the RuO_6 octahedra, and (d) the tilt angle θ of the basal O(1) plane and that of the apical O(2) at 240 K. The inset in (a) shows the ratio δ of apical bond length Ru-O(2) to planar bond length Ru-O(1) that represents the octahedral flattening [26].*

It is compelling to attribute the unusual NVTE to a mechanism where electronic correlations play a critical role. In a Mott insulator, the occurrence of orbital or magnetic ordering is always accompanied by electron localization. Localization increases kinetic energy of electrons, whereas lattice expansion reduces the kinetic energy, which is in general inversely proportional to the size of the unit cell. Meanwhile, lattice expansion costs cohesive energy (i.e., via electron-lattice interactions). The observed NVTE happens when the lattice expansion and the energy gain from the electron-electron interaction overcome the energy cost from the electron-lattice interaction and the electron localization. When the orbital and/or magnetic order takes place, the energy gain of electrons can be described in terms of the short-range coupling parameters between orbital or spin operators; namely, if we use an effective local exchange model to describe the orbital or magnetic order,

$$H = \sum_{\langle ij \rangle} \mathcal{J}_{ij} A_i A_j$$

where A_j represents a local spin or orbital moment. The effective coupling parameters $\mathcal{J}_{ij}$ are generally determined by virtual electron hopping processes. Therefore, if the lattice expansion increases $\mathcal{J}_{ij}$, the orbital and/or magnetic order can make the NVTE more energetically favorable; this is more likely in a multi-orbital system where the Coulomb repulsion U is relatively small for the following reasons: (1) virtual hopping becomes much more complicated in effective bands due to the mixture of different orbitals; and (2) SOI and crystal field effects, which are strongly affected by the lattice expansion, become comparable to U. Ca$_2$Ru$_{1-x}$M$_x$O$_4$ materials are multi-orbital systems with comparable U and SOI, as discussed in Chapter 1. While the physics that drives NTVE in these materials is yet to be fully understood, one key, unifying characteristic is their extraordinary susceptibility to the lattice degrees of freedom, which is also at the heart of novel phenomena observed in other ruthenates.

4.3 Unconventional Magnetotransport Properties in Ca$_3$Ru$_2$O$_7$

Double-layered Ca$_3$Ru$_2$O$_7$ (n = 2) [36], a sister compound of the single-layered Ca$_2$RuO$_4$ (n = 1), displays nearly every ordered state known to condensed matter physics (except for superconductivity), including conflicting hallmarks of both insulating and metallic states [1,37–41]. It is not surprising that almost all these novel phenomena arise from a high sensitivity to stimuli that couple to the lattice. After a brief survey of general physical properties of Ca$_3$Ru$_2$O$_7$, we focus on two highly unusual combinations of phenomena exhibited by this material: colossal magnetoresistivity (CMR) via avoiding magnetic polarization, and quantum oscillations occurring in a nonmetallic state.

4.3.1 Fundamental Properties of Ca$_3$Ru$_2$O$_7$

Similar to Ca$_2$RuO$_4$, Ca$_3$Ru$_2$O$_7$ adopts a severely distorted orthorhombic structure with room-temperature lattice parameters $a = 5.3720(6)$ Å, $b = 5.5305(6)$ Å, and $c = 19.572(2)$ Å [42]. The titling of the RuO$_6$ octahedra directly impacts the band structure and drives the anisotropic properties. Ca$_3$Ru$_2$O$_7$ undergoes an AFM transition at T$_N$ = 56 K while remaining metallic, and then exhibits a Mott-like transition at T$_{MI}$ = 48 K with a dramatic reduction in the conductivity as temperature decreases below T$_{MI}$ (**Figs. 4.6a–4.6c**) [1,36,39]. This transition is accompanied by an abrupt shortening of the c-axis lattice parameter below T$_{MI}$ (right scale in **Fig. 4.6b**). The resultant strong magnetoelastic coupling induces Jahn-Teller distortions of the RuO$_6$ octahedra [22,44,45], thus lowering the energy of the d_{xy}-orbitals relative to d_{zx} and d_{yz} orbitals, which defines a state of orbital order (OO), in which electrons occupy t_{2g}-orbitals in a preferred manner. This yields an orbital distribution with two electrons occupying the lower d_{xy}-orbital, and one electron occupying each of the nearly degenerate, higher-energy d_{zx} and d_{yz} orbitals. The OO state sets in at T$_{MI}$ = 48 K, which explains the localized behavior in the electrical resistivity (**Fig. 4.6b**). Indeed, Raman scattering studies of Ca$_3$Ru$_2$O$_7$ [22,43–45] indicate the OO formation below T$_{MI}$ with the opening of a charge gap, $\Delta_c \sim 0.1$ eV, and the concomitant softening and broadening of an out-of-phase O phonon mode (see Appendix, Section D).

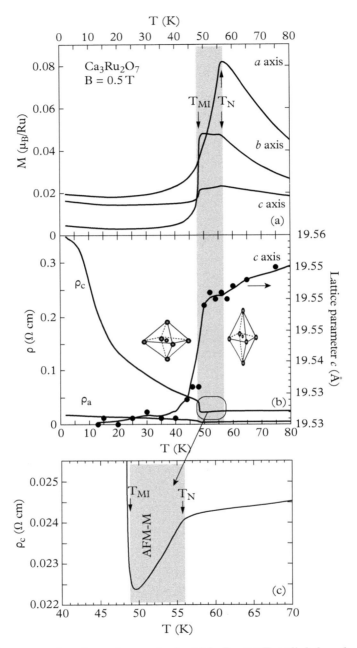

Fig. 4.6 *(a) Temperature dependence of magnetization M for B = 0.5 T applied along the a, b, and c axis, respectively. (b) Temperature dependence of the c-axis resistivity* ρ_c *(left-hand scale) and the c-axis lattice parameter (right-hand scale). (c) Enlarged* ρ_c *in the shaded area defines the antiferromagnetic metallic (AFM-M) region between* T_{MI} *and* T_N*. Note the abrupt shortening of the c axis near* T_{MP} *which flattens* RuO_6 *octahedra [39].*

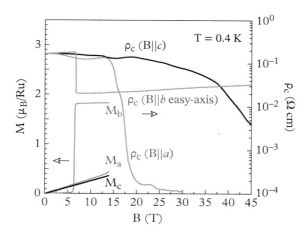

Fig. 4.7 *Magnetic field dependence of the c-axis resistivity ρ_c for B $||$ a, b, and c axes (right scale) up to 45 T, at T = 0.4 K. Isothermal magnetization M for B $||$ a, b, and c axes at T = 2 K (left scale) is plotted for comparison. Note that ρ_c for B $||$ easy b axis is greatest when B > 39 T; in contrast, ρ_c is smallest for B $||$ hard a axis.*

An antiferromagnetic metallic (AFM-M) state that is rarely seen in oxides exists between T_{MI} and T_N (see **Fig. 4.6c**) [36]. Moreover, the system is extraordinarily anisotropic, as shown in **Fig. 4.6a**. The low-field magnetization $M_b(T)$ for the *b* axis (the magnetic easy axis) exhibits two phase transitions at T_N = 56 K and T_{MI} = 48 K. In contrast, $M_a(T)$ for the *a* axis exhibits no discernable anomaly at T_{MI} but a sharp peak at T_N. For applied field parallel to the *c* axis, the two transition signatures are simultaneously observed, but they are considerably weakened (**Fig. 4.6c**).

The anisotropy of the magnetic state is also reflected in the isothermal magnetization (**Fig. 4.7**), which displays a first-order metamagnetic transition at B = 6 T applied along the easy *b* axis, leading to a spin-polarized state with a saturation moment M_s = 1.73 μ_B/Ru, which is more than 85% of the hypothetical saturation moment (2 μ_B/Ru) expected for an S = 1 system. The behavior is completely different if the field is applied along the *a* or *c* axis, in part due to a strong anisotropy field of 22.4 T [46]. The rich, anisotropic phases illustrated in **Fig. 4.7** are the central topic of the following discussion.

4.3.2 Colossal Magnetoresistivity via Avoiding a Spin-Polarized State

Conventional CMR occurs only within a spin-polarized state with magnetic field applied along the magnetic easy axis. Ca$_3$Ru$_2$O$_7$ is fundamentally different from all other CMR materials in that it exhibits CMR only when the spin-polarized state is *avoided*.

Figure 4.7 shows the magnetic field dependence of the *c*-axis resistivity ρ_c for B$||$a, b, and c axes at T = 0.4 K and 0 ≤ B ≤ 45 T. ρ_c is extraordinarily sensitive to the orientation of B. For $B||b$ axis (magnetic easy axis), ρ_c shows an abrupt drop by an order of magnitude at 6 T; this is identified with a first-order metamagnetic transition leading to a spin-polarized state with the saturated moment M_s, of 1.73 μ_B/Ru, as discussed earlier (see M_b in **Fig. 4.7**). The reduction of ρ_c is attributed to an increase of coherent motion of the electrons between Ru-O planes separated by insulating Ca-O planes; this behavior recalls "spin-filters," where the probability of tunneling depends on the angle between the spin magnetizations of adjacent FM layers. The data in **Fig. 4.7** indicate that a fully spin-polarized state can lower the resistivity by at most a factor of 10. As B is further increased from 6 to 45 T, ρ_c increases *linearly with B* by more than 30% (ρ_c in right scale in **Fig. 4.7**); this is particularly intriguing since a quadratic dependence is expected for simple metals [47,48]. It is worth mentioning that a non-saturating positive magnetoresistance is a hallmark of many of topological semimetals where charge compensation plays an important role in determining magnetotransport properties [49,50]. There is no metamagnetic transition for $B||a$ axis (magnetic hard axis), and the system remains AFM (see M_a in **Fig. 4.7**). In sharp contrast to ρ_c for $B||b$ axis, ρ_c for $B||a$ rapidly decreases by as much as three orders of magnitude at a critical field, $B_c = 15$ T, which is two orders of magnitude more than that for $B||b$ axis, where spins are near saturation above B_c. On the other hand, ρ_c for $B||c$ axis displays Shubnikov-deHaas (SdH) oscillations with low frequencies of 28 T and 10 T [41]. Remarkably, ρ_c for $B||c$ axis is much smaller for B > 39 T than ρ_c for $B||b$ axis, the magnetic easy axis (**Fig. 4.7**).

The nearly fully polarized state for $B_{||b} > 6$ T only reduces resistivity ρ by one order of magnitude; therefore it cannot account for the observed three orders of magnitude decrease in ρ_c when $B_{||a} > 15$ T. This indicates that the spin degree of freedom alone cannot account for the behavior observed in **Fig. 4.7**. This is highly unusual because a fully spin-polarized state is essential for minimal magnetoresistance in all other magnetoresistive materials [51].

Complementary evidence for the evolution of the field-induced magnetic and orbital phases inferred from the magnetic and transport behavior (**Fig. 4.7**) is provided by observations of rapid changes of the Ru-O phonon frequency with applied magnetic field seen in Raman spectra (see Appendix, Section D), as shown in **Fig. 4.8** [44]. Application of magnetic field along the a axis clearly favors the CMR, whereas applying $B||b$ axis generates a rich phase diagram. Below 40 K, applied magnetic field drives the system from an AFM/OO to an FM/OO state. The evolution of the magnetic/orbital configuration is associated with the Jahn-Teller effect, which appears to be strongest in the vicinity of T_{MI}.

In addition, the temperature dependence of ρ_c for $B||a$ and b axis is displayed in **Fig. 4.9**. For $B||b$ axis and at low temperatures, ρ_c decreases abruptly by about an order of magnitude when $B \geq 6$ T, at which the first-order metamagnetic transition leads to the spin-polarized state, as seen in **Fig. 4.7**. A further increase in B results in only slightly higher resistivity at low temperatures. In fact, ρ_c at B = 28 T still exhibits nonmetallic behavior (**Fig. 4.9a**). In sharp contrast, when $B||a$ axis, T_{MI} decreases systematically at

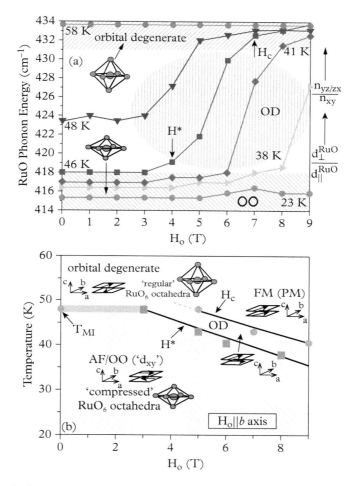

Fig. 4.8 *(a) Ru-O phonon energy vs. magnetic field for various temperatures with H$_o$ || b axis, illustrating phases of orbital order (OO) and orbital disorder (OD). (b) H$_o$-T phase diagram with H$_o$ || b axis. Squares denote a transition between AF/OO and OD phases, and circles denote transitions between OD and FM/OO phases. Note that RuO$_6$ octahedra are compressed below T$_{MI}$ = 48 K [44].*

a rate of 2K/T, and disappears at B > 24 T, as seen in **Fig. 4.9b**. These results once again suggest a strong magneto-orbital coupling. It is also striking that at B = 30 T, where the metallic state is recovered and AFM order is completely suppressed (**Fig. 4.9c**), $\rho_c \sim T^{1.2}$ for T < 56 K [39]. The non-integral, near-linear power law may indicate a proximity to a quantum critical point.

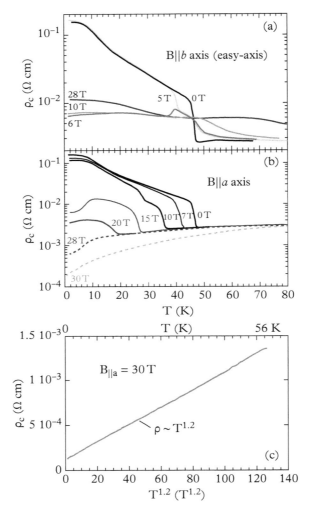

Fig. 4.9 *Temperature dependence of c-axis resistivity ρ_c at representative magnetic fields B applied along (a) the b axis and (b) a axis for 1.2 < T < 80 K. Note that T_{MI} is effectively suppressed by applied field and eventually vanishes for B > 24 T when B $\parallel$ a. (c) ρ_c for $B_{\parallel a}$ = 30 T. Note the $T^{1.2}$-dependence implies the proximity of a quantum critical point [39].*

4.3.3 Shubnikov-de Haas Effect in the Nonmetallic State

Quantum oscillations are not expected in nonmetallic materials, but nevertheless occur in $Ca_3Ru_2O_7$ [37,41]. As shown in **Fig. 4.10a**, for B$\parallel$ c axis, ρ_c displays slow yet strong Shubnikov-de Haas (SdH) oscillations, signaling the existence of very small Fermi surface cross-sectional areas. It is plausible that the d_{xy}-orbitals give rise to small lens-shaped

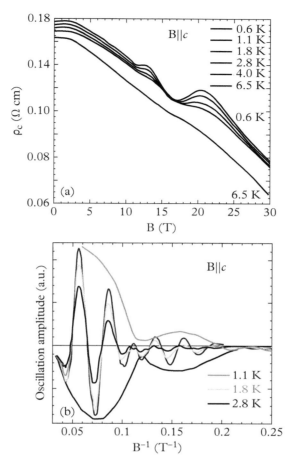

Fig. 4.10 *(a) Magnetic field dependence of ρ_c at a few representative temperatures up to 30 T applied along the c axis; (b) the corresponding oscillation amplitude as a function of 1/B. The estimate frequencies are 10 and 28 T, signaling extremely small Fermi surface [37].*

Fermi surface pockets that are very sensitive to slight structural changes. The observed oscillations must then be associated with the motion of the electrons in the *ab* plane—that is, with the d_{xy}-orbitals. As seen in **Fig. 4.10b**, the oscillations in ρ_c correspond to two extremely low frequencies, $f_1 = 28$ T and $f_2 = 10$ T, which, based on crystallographic data and the Onsager relation $F_0 = A(h/4\pi^2 e)$ (*e* is the electron charge, *h* Planck's constant), correspond to a cross-sectional area A of only 0.2% of the first Brillouin zone. From the temperature dependence of the SdH amplitude, the cyclotron effective mass is estimated to be $\mu_c = (0.85 \pm 0.05)m_e$, where m_e is the free-electron mass. This is markedly smaller than the thermodynamic effective mass ($m^* \sim 3m_e$) estimated from the electronic coefficient γ from specific heat data. There are three possible sources for this discrepancy: (1) The cyclotron effective mass is measured in a large magnetic field that quenches correlations, while the specific heat is a zero-field measurement. (2) μ_c refers to only one closed orbit, whereas the thermodynamic effective mass reflects an average over the entire Fermi surface. (3) The Dingle temperature, $T_D = h/4\pi^2 k_B \tau$, is a measure of scattering at the Fermi surface (estimated to be 3 K for the observed oscillations, comparable to those of good organic metals), and the Dingle temperatures of the highest mass electron orbits may be too large to permit resolved oscillations.

Quantum oscillations are also observed in the bc plane, or when B rotates toward the b axis, but not in the ac plane [41], reinforcing the evidence for the highly anisotropic nature of the electronic structure of $Ca_3Ru_2O_7$.

4.3.4 Oscillatory Resistivity Periodic in B

Oscillatory resistivity periodic in B is not consistent with SdH oscillations, which are periodic in 1/B, as shown in **Fig. 4.10a**. However, for B$||$[110], ρ_c exhibits magneto-

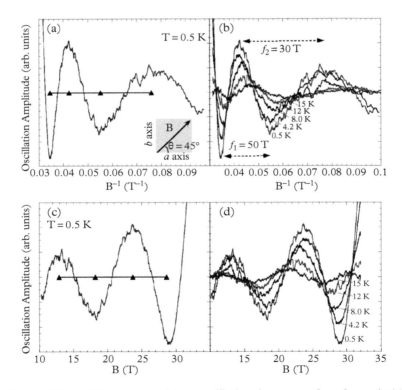

Fig. 4.11 *The amplitude of the magnetoresistance oscillations for current along the c axis: (a) as a function of B^{-1} for B $||$ [110] and T = 0.5 K; and (b) for various temperatures up to 15 K (note oscillations are not periodic in 1/B). (c) The amplitude of the oscillations as a function of B for B $||$ [110] and T = 0.5 K; and (d) amplitudes for various temperatures up to 15 K (note oscillations are periodic in B) [41].*

resistance oscillations; these oscillations are periodic in B with a period $\Delta B = 11$ T, and are persistent up to T = 15 K, as shown in **Fig. 4.11** [41]. This is corroborated by plotting the data, both as a function of 1/B (**Figs. 4.11a** and **4.11b**) and B (**Figs. 4.11c** and **4.11d**). The oscillations die off rapidly if B slightly departs (e.g., by $\pm 5°$) from the [110] direction.

Due to the Pauli principle, electrons are constrained to move about the Fermi surface at temperatures well below the Fermi energy. Magnetoresistance oscillations periodic in 1/B (i.e., the SdH effect) are a manifestation of the constructive interference of quantized extremal orbits on the Fermi surface with cross-sections perpendicular to the applied magnetic field, as discussed in Section 4.3.3. A projection of the real-space trajectory of a nearly free electron onto a plane perpendicular to B reproduces the k-space trajectory, but rotated by $\pi/2$ and scaled by a factor, $c\hbar/|e|B$. Hence, trajectories with constructive interference in real space are expected to be periodic in B rather than 1/B (the frequency is proportional to the cross-sectional area in reciprocal space, so that the relation to the real space is B^2). Oscillations in the magnetoresistivity periodic in B are realized in some mesoscopic systems where they are always related to finite-size effects. Examples are (1) the Aharanov-Bohm (AB) effect [48], (2) the Sondheimer effect [48,52], and (3) edge states in quantum dots [53]. Note that each of the cases involves a geometrical confinement of charge carriers.

The AB interference occurs when a magnetic flux threading a metallic loop changes the phase of the electrons, generating oscillations in the magnetoresistance, and is observed only in mesoscopic conductors—not in bulk materials. The Sondheimer effect requires a thin metallic film with the wavefunction vanishing at the two surfaces. The thickness of the film has to be comparable with the mean free path. This gives rise to boundary scattering of the carriers that alters the free-electron trajectories and the conditions for interference. Finally, quantum Hall edge states require real-space confinement [53].

Since a bulk, three-dimensional material has no real-space confinement for the orbits of the carriers, the most likely explanation for the periodicity as a function of B is a Fermi surface cross-section that changes with field. The t_{2g} orbitals have off-diagonal matrix elements arising from the orbital Zeeman effect, and hence couple directly to the magnetic field. It is therefore conceivable that a changing magnetic field could lead to a dramatic change of the Fermi surface if it points along a special direction. Note that the Fermi surface pockets involved are very small (low frequencies as a function of 1/B) and susceptible to external influences. If there is more than one conducting portion of the Fermi surface, occupied states can be transferred from one pocket to another with relatively small changes in the external parameters. We also must note that the AB effect at finite temperature would show the same amplitude dependence [54,55]. It is still perplexing that the cross-section of the observed pocket is only 0.2% of the Brillouin zone, so the position of the Fermi energy is fixed at the non-quantized level of other Fermi surface branches. In such a situation, the density of states oscillates only against 1/B. In addition, if the origin of the oscillations periodic in B is ascribed to the Landau quantization, it is then perplexing as to why there are no SdH oscillations for B||[110], simultaneous with the oscillations periodic in B.

The observations of strong oscillations in the magnetoresistance of $Ca_3Ru_2O_7$ (periodic both in B and 1/B) certainly reflect a high sensitivity of the quantized electron orbits on the orientation of B. These novel phenomena strongly suggest that there is a critical role played by the lattice and orbital degrees of freedom via a coupling of the t_{2g} orbitals to the magnetic field, and certainly merit more experimental and theoretical attention.

4.3.5 Summary Remarks

The unusual and sometimes spectacular physical properties of $Ca_3Ru_2O_7$ reflect most of the ordered states known in condensed matter physics, and also include unusual phenomena not found in other materials. It has become increasingly clear that the lattice and orbital degrees of freedom drive the complex phase diagram of $Ca_3Ru_2O_7$ via coupling of the orbital degrees of freedom to the electron spins (spin-orbit interactions), and also, via a Jahn-Teller effect, to lattice degrees of freedom. These phenomena present profound intellectual challenges and pose a set of intriguing questions whose answers will eventually establish a deeper understanding of highly correlated electron behavior in transition-metal oxides.

4.4 Pressure-Induced Transition from Interlayer Ferromagnetism to Intralayer Antiferromagnetism in $Sr_4Ru_3O_{10}$

The $Sr_{n+1}Ru_nO_{3n+1}$ compounds are metallic and tend to be FM, as discussed in Section 1.5. However, the triple-layered $Sr_4Ru_3O_{10}$ (n = 3) [56] is precariously positioned on the borderline separating the ferromagnet $SrRuO_3$ (n = ∞) [57] and the field-induced metamagnet $Sr_3Ru_2O_7$ (n = 2) [58,59], and displays complex phenomena, including tunneling magnetoresistance and quantum oscillations [38,60,61], as well as a switching effect [62]. However, the most distinct, intriguing hallmark of $Sr_4Ru_3O_{10}$ is its seemingly contradictory magnetic behavior: for an external magnetic field applied along the c axis (perpendicular to the layers), it exhibits itinerant FM with a Curie temperature T_C = 105 K and a saturation moment greater than 1.0 μ_B/Ru (**Fig. 4.12**). On the other hand, when the magnetic field is applied within the basal plane, a pronounced peak in the magnetization appears near T_M = 50 K (**Fig. 4.12a**) and a sharp metamagnetic transition is evident near H_C = 2.5 T (**Fig. 4.12b**) [38,56,60,61,63,64]. In particular, the heat capacity data in **Fig. 4.13** clearly confirm the existence of a robust metamagnetic transition in this mostly ferromagnetic system. This situation recalls the metamagnetic transitions out of Stoner exchange-enhanced paramagnetism in Sr-based Ruddlesen-Popper compounds. The coexistence of the interlayer FM and the intralayer metamagnetism is not anticipated from simple theoretical arguments [65–69]. A two-dimensional, tight-binding electron gas has a logarithmic divergence in the density of states which, depending on the position of the Fermi level, can yield FM, metamagnetism, and a quantum critical point by varying applied pressure [69].

The peculiar behavior of $Sr_4Ru_3O_{10}$ has drawn considerable attention in recent years [64,70–72]. Discussions have mainly focused on the relationship between the interlayer FM and the intralayer metamagnetism below T_M, the onset of the metamagnetic state.

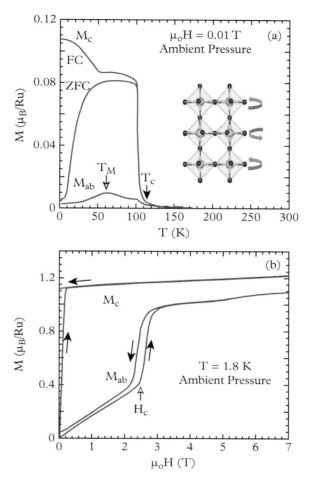

Fig. 4.12 *(a) Temperature dependence of the magnetization M for $Sr_4Ru_3O_{10}$ at $\mu_oH = 0.01$ T, and (b) isothermal magnetization M(H) at T = 1.8 K for both the basal-plane M_{ab} and the c-axis M_c at ambient pressure. Inset: Schematic of the triple-layered crystal structure; the curved arrows indicate the rotation of the RuO_6 octahedra [38].*

A Raman study explores the spin-lattice coupling as a function of temperature, magnetic field, and pressure, and informs a microscopic description of the structural and magnetic phases [45]. Specifically, magnetic-field-induced changes in the phonon spectra reveal spin-reorientation transitions and strong magnetoelastic coupling below T_M. In addition, a rapid increase in the c-axis lattice parameter below T_M at ambient conditions is observed [73], signaling a critical role for strong spin-lattice coupling in determining the magnetic state. A similar conclusion is drawn from studies of Raman scattering [45] and Ca- and La-doping behavior of $Sr_4Ru_3O_{10}$ [61]. A magnetic and transport study of

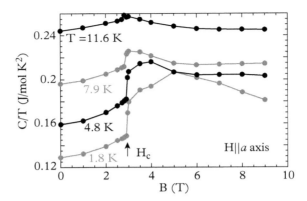

Fig. 4.13 *Magnetic field dependence of the heat capacity divided by temperature C/T for a few representative temperatures. The data confirm that the metamagnetization transition at the critical magnetic field $H_C = 2.5$ T is a bulk and robust feature of $Sr_4Ru_3O_{10}$.*

nanoscale flakes (30–350 nm) of $Sr_4Ru_3O_{10}$ revealed a drastic thickness effect in the evolution from c-axis FM to a basal-plane AFM state [74]. In short, it is increasingly clear that spin-lattice coupling is critical to the apparent coexistence of the interlayer FM and the intralayer metamagnetism.

Application of pressure is a powerful, fundamental tool for tuning lattice properties without introducing impurities, and provides insights into the complex behavior of $Sr_4Ru_3O_{10}$ [75]. Application of pressure induces a rapid evolution from c-axis-itinerant FM at ambient pressure to basal-plane-itinerant AFM at pressures near 25 kbar, accompanied by a diminishing magnetic anisotropy that promotes the metamagnetic state under ambient conditions.

$Sr_4Ru_3O_{10}$ has an orthorhombic structure with a *Pbam* space group and room-temperature lattice parameters $a = 5.4982$ Å, $b = 5.4995$ Å, and $c = 28.5956$ Å [56]. One important structural detail is that in the outer two perovskite layers, the RuO_6 octahedra are rotated 5.25° about the c axis, whereas in the middle layer they are rotated 10.6° about the c axis in the opposite direction (see **Fig. 4.12a**, inset). This structural feature has significant implications for the spin configuration and physical properties discussed later, due to the action of strong spin-lattice coupling.

The spins are ferromagnetically aligned along the easy c axis in $Sr_4Ru_3O_{10}$ at ambient conditions, effectively forming FM chains [45]. Application of a modest pressure readily destabilizes the c-axis FM state and fosters emergent AFM correlations with spins primarily aligned within the basal plane below T_M, as illustrated in **Fig. 4.14** [75]. The pressure-induced change in T_M is clearly identified by the shift of a corresponding peak in the basal-plane magnetization M_{ab} and indicates an astonishing fourfold enhancement at 10 kbar (**Fig. 4.14a**), whereas the c-axis magnetization M_c undergoes a comparable reduction (**Fig. 4.14b**). A peak in M_c emerges at around $P = 8$ kbar and becomes well-defined at $P = 10$ kbar. The occurrence of peaks in both M_{ab} and M_c at $P = 10$ kbar indicate an AFM ordered state is emerging (**Fig. 4.14c**). Note that the magnetic anisotropy $M_c/M_{ab} \sim 10$

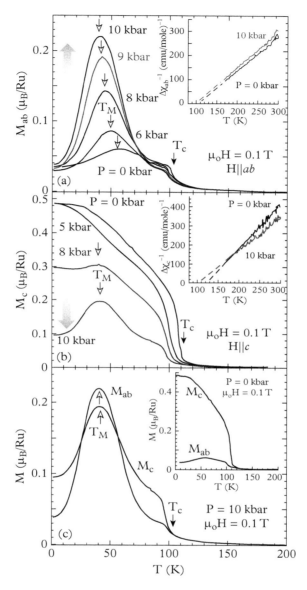

Fig. 4.14 *Temperature dependence of the magnetization M at $\mu_o H = 0.1$ T for (a) the basal plane M_{ab}, (b) the c-axis M_c at representative pressures, and (c) M_{ab} and M_c at P = 10 kbar for comparison. Insets in (a) and (b): $1/\Delta\chi$ at P = 0 and 10 kbar vs. T, where $\Delta\chi$ = magnetic susceptibility χ − Pauli susceptibility χ_o, for the basal plane and the c axis, respectively. Inset in (c): M_{ab} and M_c at P = 0 kbar [75].*

for P = 0 kbar and T = 50 K (**Fig. 4.14c,** inset) approaches unity near P = 10 kbar. In addition, the signature step in M_c at T_C decreases with P. This is consistent with a decrease in the Curie-Weiss temperature θ_{cw} under pressure, which is illustrated in plots of $1/\Delta\chi$ ($\Delta\chi$ = magnetic susceptibility χ-Pauli susceptibility χ_o) for the basal plane and the c axis, respectively (**Figs. 4.14a** and **4.14b,** insets). As seen, θ_{cw} (intercept on the horizontal axis) changes from 122 K (105 K) at P = 0 kbar to 100 K (98 K) at 10 kbar for the c axis (basal plane), suggesting that application of pressure alters the exchange interaction. Overall, applied pressure rapidly drives the magnetic state from a c-axis-FM state toward a basal-plane-AFM state below T_M. The fact that a modest 10-kbar pressure can cause such drastic changes in the magnetization and its anisotropy demonstrates an unusually strong magnetoelastic effect is at play in this material [75].

The Raman study of $Sr_4Ru_3O_{10}$ revealed a strong spin-lattice coupling of 5.2 cm^{-1} below T_M [45]. The pressure-induced AFM state is likely a result of a flattening of the RuO_6 octahedra, in which d_{xy}-orbitals have the lowest energy and are fully occupied, whereas the d_{xz} and d_{yz} orbitals are half filled. Superexchange interactions mediated by electrons in the d_{xz} and d_{yz} orbitals favor an AFM state, according to electronic band structure calculations [76,77]. This scenario also explains the observation of an unusual increase in the c-axis resistivity ρ_c with P at T > T_C, as shown in **Fig. 4.15**. The basal-plane resistivity ρ_{ab} exhibits only a small decrease with P at higher temperatures, and presumably results from band broadening (**Fig. 4.15a**). The slight change in ρ_{ab} implies that soft-phonon and spin-disorder scattering within the basal plane are largely unaffected under the conditions studied. In sharp contrast, the c-axis resistivity ρ_c increases by a factor of two at T > T_C (**Fig. 4.15b**), and the ratio of ρ_c/ρ_{ab} at 300 K rises from 3.8 at ambient pressure to 7.2 at 25 kbar (**Fig. 4.15a,** inset).

Although a structural study of $Sr_4Ru_3O_{10}$ under pressure is not yet available for a more conclusive discussion, the octahedral tilt affects the orbital overlap, and qualitative and quantitative changes in resistivity are possible. The behavior demonstrated in **Fig. 4.15** seems to suggest that applied pressure induces a tilting of the RuO_6 octahedra, which reduces the Ru-O-Ru bond angle from 180°, which in turn reduces the overlap of p- and d-orbitals. The reduced overlap might help explain the significant increase in ρ_c for T > T_C (**Fig. 4.15b**), and is also consistent with an anomalous pressure dependence of the 380 cm^{-1} B_{1g} phonon mode at low temperatures, which is attributed to a buckling of the RuO_6 octahedra [45]. Interestingly, the magnitude of ρ_c for T < T_M is much less affected by applied pressure (**Fig. 4.15b**). The rapidly reduced ρ_c below T_M signifies a strengthened overlap of d_{xz}/d_{yz} orbitals and, more importantly, the existence of long-range magnetic order at P = 25 kbar. Overall, electrical transport is intimately coupled to magnetism and spin-disorder scattering; and long-range order significantly reduces both phonon and spin scattering.

Also of interest is the temperature dependence of $\rho_c \sim T^\alpha$ at low temperatures (1.8–10 K), where the exponent α changes significantly from near 2 at ambient pressure to 3/2 (**Fig. 4.15b, inset**), which suggests a dominance of AFM spin fluctuations and a breakdown of the Fermi liquid model [65]. As shown in **Fig. 4.15c**, α changes more rapidly near P = 10 kbar, which may mark an onset of a more isotropic, itinerant AFM state. The change in α is correlated with changes of T_M, which increases at P ≥ 10 kbar, whereas T_C steadily decreases over the same range (**Fig. 4.15b** shows ρ_c at a few representative

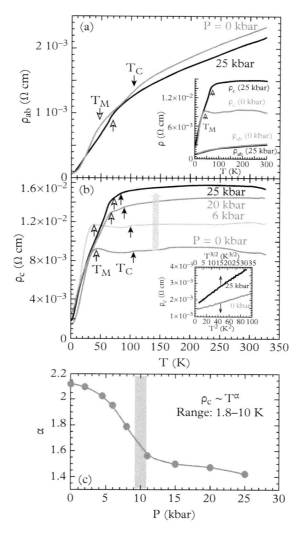

Fig. 4.15 *Temperature dependence of the electrical resistivity for (a) the basal-plane ρ_{ab} and (b) the c-axis ρ_c at representative pressures. Note the modest decrease in ρ_{ab} and the considerable increase in ρ_c with P, which is marked by the broad arrow. (c) The exponent α of $\rho_c \sim T^\alpha$ as a function of pressure P. Note that the shaded area marks a rapid change in α [75]. Inset in (a): A comparison between ρ_{ab} and ρ_c at P = 0 and 25 kbar. Inset in (b): vs T^2 at P = 0 and $T^{3/2}$ (upper horizontal axis) at P = 25 kbar .*

pressures). The opposite response of T_M and T_C to pressure further confirms that the basal-plane-AFM state becomes more energetically favorable than the c-axis-FM state with an increasing distortion of the RuO_6 octahedra with increasing P. The data in **Fig. 4.14** imply that the emergent basal-plane-AFM state is dominant at P ≥ 10 kbar.

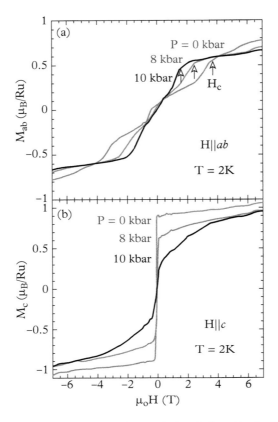

Fig. 4.16 *Isothermal magnetization at T = 2 K for (a) the basal-plane M_{ab} for H $||$ basal plane and (b) the c-axis M_c for H $||$ c axis at a few representative pressures [75].*

The evolution of both the basal-plane and c-axis isothermal magnetizations at low temperatures also indicates that applied pressure enhances M_{ab} and weakens M_c, as shown in **Fig. 4.16**. The critical field of the metamagnetic transition, H_c, decreases with increasing P, indicative of an increasingly weakened magnetic anisotropy (**Fig. 4.16a**). Note that magnetic field applied along the c axis helps elongate the RuO_6 octahedra along the c axis, which enhances M_c. This effect competes with applied pressure, which tends to compress the RuO_6 octahedra. The result of the two competing effects may explain why M_c changes only modestly with P (**Fig. 4.16b**). Moreover, both M_{ab} and M_c exhibit a sizable hysteresis effect at ambient pressure (**Fig. 4.12**), but this hysteresis almost vanishes at P = 10 kbar (not shown), consistent with a weakened FM state.

$Sr_4Ru_3O_{10}$ exhibits a large negative magnetoresistivity $\rho(H)$ with an overall reduction of up to 80% (**Fig. 4.17**). The basal-plane resistivity $\rho_{ab}(H)$ rises initially due to the canted AFM state, and then drops abruptly when H is strong enough to align the spins in a collinear fashion, which reduces spin scattering. Specifically, ρ_{ab} shows two peaks near two critical fields, H_c and H_{c2}, respectively. The peak at H_c marks the metamagnetic

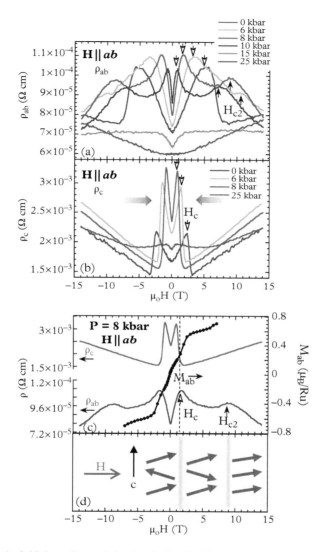

Fig. 4.17 *Magnetic-field dependence of the electrical resistivity at 2 K for (a) the basal plane* ρ_{ab}, *(b) the c-axis* ρ_c *at representative pressures, and (c)* ρ_{ab} *and* ρ_c *and* M_{ab} *(right scale) at P = 8 kbar for comparison. (d) Schematic for field-induced changes in the spin configuration. Note that the two broad horizontal arrows in* **Fig. 4.17b** *are to highlight the rapid decrease in* H_c *with P, and the vertical dashed line in* **Fig. 4.17c** *indicates* H_c *[75].*

transition that indicates a spin-flip (presumably in the middle layer) that aligns the middle layer spins with those in the outer layers; therefore all spins, although canted, are approximately collinear along with the direction of H, which reduces spin scattering, and explains the abrupt drop in ρ_{ab} near H_c (**Fig. 4.17a**). This is then followed at higher fields by another drop in ρ_{ab} at H_{c2} (**Figs. 4.17a** and **4.17c**). It indicates an additional spin alignment that eventually diminishes the spin canting and further reduces spin scattering. The evolution of the spin configuration with H is schematically illustrated in **Fig. 4.17d**. Note that ρ_c drops even more sharply near H_c but exhibits no anomaly at H_{c2}; instead it increases linearly with H when $H > H_c$ (**Figs. 4.17b** and **4.17c**). Since the direction of applied magnetic field H is perpendicular to the direction of electrical current, the linear rise of ρ_c with H could be a result of the deflection of electrons by the Lorentz force (orbital magnetoresistance). On the other hand, this linear field-dependence is strikingly similar to that observed in $Ca_3Ru_2O_7$ (see **Fig. 4.7**), which is attributed to an orbital order that strengthens with increasing magnetic field, which hinders electron hopping, as discussed in Section 4.3. The two critical fields H_c and H_{c2} rapidly decrease with increasing P, and eventually vanish at $P = 25$ kbar, where $\rho_{ab} \sim H^2$ (ρ_c behaves similarly above 1 T).

It needs to be pointed out that the field dependences of both ρ_{ab} and ρ_c bear a strong resemblance to the bulk spin-valve effect observed in bilayered $Ca_3(Ru_{1-x}Cr_x)_2O_7$, which originates from inhomogeneous exchange coupling and soft and hard bilayers having antiparallel spin alignments [78]. Consistently, the distinct behavior demonstrated by ρ_{ab} and ρ_c (**Figs. 4.17a–4.17c**) implies a strong exchange anisotropy that must arise from the competition between FM and AFM correlations. The anisotropy and spin-valve effects are clearly highly susceptible to applied pressure, and eventually vanish when P exceeds 25 kbar.

A temperature-pressure phase diagram for $Sr_4Ru_3O_{10}$ can be generated using current results, as shown in **Fig. 4.18**. The lattice distortions destabilize the c-axis-FM state, and foster a basal-plane-AFM state. As a result, T_C decreases at a rate $dT/dP \approx -1$ K/kbar;

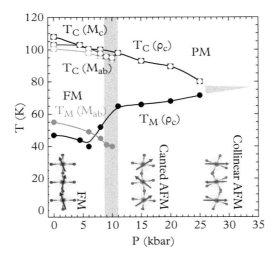

Fig. 4.18 *A T-P phase diagram generated based on the magnetic and transport results. Note that T_C and T_M appear to merge as the antiferromagnetic state is fully developed at $P \geq 25$ kbar. Note that the shaded areas near 10 kbar mark a crossover from the c-axis-FM state to a predominately basal-plane-AFM state [75].*

whereas T_M, which defines the onset of the AFM state, decreases initially and then rises for P ≥ 10 kbar. Indeed, the rapid change in the temperature dependence of ρ_c (and ρ_{ab}) from T^2 to $T^{3/2}$ below 10 K and near P = 10 kbar marks a crossover from the *c*-axis-FM state to a predominantly basal-plane-AFM state. Nevertheless, the opposite pressure responses of T_C and T_M point to a merging of the two magnetic states at P ≥ 25 kbar, at which a collinear, itinerant AFM state is presumably fully established. The existence of this pressure-induced long-range order at 25 kbar is clearly indicated by the abrupt drop in ρ_c below T_M. The attainment of a collinear antiferromagnetic state out of a low-pressure canted state should also generate a strongly varying anomalous (topological) Hall effect, due to variations in the scalar spin chirality [79].

It is remarkable that the FM state with T_C = 165 K in the sister compound $SrRuO_3$ decreases with pressure at a slower rate dT/dP ≈ −0.68 K/kbar, and vanishes only in a much higher pressure range of 170–340 kbar, where a paramagnetic state emerges [80]. This sharply contrasts with the high tunability offered by pressure applied in $Sr_4Ru_3O_{10}$ and once again highlights the critical role the lattice plays in these materials. This point is also corroborated by the observation that the magnetic configuration sensitively changes with the sample thickness (in nanoscale) in $Sr_4Ru_3O_{10}$ [74,81].

4.5 General Remarks

This chapter presents the NVTE via orbital and magnetic orders, CMR via avoiding magnetic polarization, quantum oscillations in a nonmetallic state, oscillatory resistivity periodic in B, and the transition from the interplane FM to the intraplane AFM order under modest pressures. These are only a glimpse into a panoply of exotic states exhibited by this class of materials. However, at the heart of all these novel phenomena is an extraordinary sensitivity of these materials to the lattice and orbital degrees of freedom. This sensitivity results from a delicate interplay between all fundamental interaction energies. Of primary importance is the enhanced SOI (proportional to the square of the atomic number Z^2), the extended 4*d*-orbitals leading to a reduced relative strength of the electron-electron repulsion (U from the Hund's rule coupling), and increased *d-p*-orbital hybridization. The comparable strength of these basic interactions is a central, dominating characteristic of 4*d*- and 5*d*-transition metal oxides.

Further Reading

- J.M.D. Coey. *Magnetism and Magnetic Materials.* Cambridge (2018)
- Stephen Blundell. *Magnetism in Condensed Matter.* Oxford (2001)
- Yoshinori Tokura. *Colossal Magnetoresistive Oxide.* Gordon and Beach Science Publishers, Australia (2000)
- Gang Cao and Lance E. DeLong, Eds. *Frontiers of 4d- and 5d-Transition Metal Oxides.* World Scientific (2013)

References

1. Cao, G., DeLong, L., (Eds.) *Frontiers of 4d- and 5d-transition Metal Oxides* (World Scientific, 2013)
2. Maeno, Y., Hashingmoto, H., Yoshida, K., Ishizaki, S., Fujita, T., Bednorz, J.G., Lichtenberg, F., Superconductivity in a layered perovskite without copper, *Nature.* **372**, 532 (1994)
3. Mackenzie, A.P., Maeno, Y., The superconductivity of Sr_2RuO_4 and the physics of spin-triplet pairing, *Rev. Mod. Phys.* **75**, 657 (2003)
4. Hicks, C.W., Brodsky, D.O., Yelland, E.A., Gibbs, A. S., Bruin, J.A.N., Barber, M.E., Edkins, S.D., Nishimura, K., Yonezawa, S., Maeno, Y., Mackenzie, A. P., Strong Increase of T_c of Sr_2RuO_4 Under Both Tensile and Compressive Strain, *Science.* **344**, 283 (2014)
5. Steppke, A., Zhao, L., Barber, M.E., Scaffidi, T., Jerzembeck, F., Rosner, H., Gibbs, A.S., Maeno, Y., Simon, S.H., Mackenzie, A.P., Hicks, C.W., Strong peak in T_c of Sr_2RuO_4 under uniaxial pressure, *Science.* **355**, 148 (2017)
6. Pustogow, A., Luo, Y., Chronister, A., Su, Y.-S., Sokolov, D.A., Jerzembeck, F., Mackenzie, A.P., Hicks, C.W., Kikugawa, N., Raghu, S., Bauer, E.D., Brown, S.E., Constraints on the superconducting order parameter in Sr_2RuO_4 from oxygen-17 nuclear magnetic resonance, *Nature.* **574**, 72 (2019)
7. Ishida, K., Mukuda, H., Kitaoka, Y., Asayama, K., Mao, Z. Q., Mori, Y., Maeno, Y., Spin-triplet superconductivity in Sr_2RuO_4 identified by ^{17}O Knight shift, *Nature.* **396**, 658 (1998)
8. Cao, G., McCall, S., Shepard, M., Crow, J.E., Guertin, R.P., Magnetic and transport properties of single-crystal Ca_2RuO_4: Relationship to superconducting Sr_2RuO_4, *Phys. Rev. B.* **56**, R2916(R) (1997)
9. Alexander, C.S., Cao, G., Dobrosavljevic, V., McCall, S., Crow, J.E., Lochner, E., Guertin, R.P., Destruction of the Mott insulating ground state of Ca_2RuO_4 by a structural transition, *Phys. Rev. B.* **60**, R8422(R) (1999)
10. Braden, M., André, G., Nakatsuji, S., Maeno, Y., Crystal and magnetic structure of Ca_2RuO_4: Magnetoelastic coupling and the metal-insulator transition, *Phys. Rev. B.* **58**, 847 (1998)
11. Jung, J.H., Fang, Z., He, J.P., Kaneko, Y., Okimoto, Y., Tokura, Y., Change of Electronic Structure in Ca_2RuO_4 Induced by Orbital Ordering, *Phys. Rev. Lett.* **91**, 056403 (2003)
12. Fang, Z., Terakura, K., Magnetic phase diagram of $Ca_{2-x}Sr_xRuO_4$ governed by structural distortions, *Phys. Rev. B.* **64**, 020509(R) (2001)
13. Hotta, T., Dagotto, E., Prediction of Orbital Ordering in Single-Layered Ruthenates, *Phys. Rev. Lett.* **88**, 017201 (2001)
14. Lee, J.S., Lee, Y.S., Noh, T.W., Oh, S.-J., Yu, J., Nakatsuji, S., Fukazawa, H., Maeno, Y., Electron and Orbital Correlations in $Ca_{2-x}Sr_xRuO_4$ Probed by Optical Spectroscopy, *Phys. Rev. Lett.* **89**, 257402 (2002)
15. Gorelov, E., Karolak, M., Wehling, T.O., Lechermann, F., Lichtenstein, A.I., Pavarini, E., Nature of the Mott Transition in Ca_2RuO_4, *Phys. Rev. Lett.* **104**, 226401 (2010)
16. Liu, G.-Q., Spin-orbit coupling induced Mott transition in $Ca_{2-x}Sr_xRuO_4$ ($0 \leq x \leq 0.2$), *Phys. Rev. B.* **84**, 235136 (2011)
17. Cao, G., McCall, S., Dobrosavljevic, V., Alexander, C.S., Crow, J.E., Guertin, R.P., Ground-state instability of the Mott insulator Ca_2RuO_4: Impact of slight La doping on the metal-insulator transition and magnetic ordering, *Phys. Rev. B.* **61**, R5053(R) (2000)
18. Nakamura, F., Goko, T., Ito, M., Fujita, T., Nakatsuji, S., Fukazawa, H., Maeno, Y., Alireza, P., Forsythe, D., Julian, S.R., From Mott insulator to ferromagnetic metal: A pressure study of Ca_2RuO_4, *Phys. Rev. B.* **65**, 220402(R) (2002)

19. Zegkinoglou, I., Strempfer, J., Nelson, C.S., Hill, J.P., Chakhalian, J., Bernhard, C., Lang, J.C., Srajer, G., Fukazawa, H., Nakatsuji, S., Maeno, Y., Keimer, B., Orbital Ordering Transition in Ca_2RuO_4 Observed with Resonant X-Ray Diffraction, *Phys. Rev. Lett.* **95**, 136401 (2005)

20. Nakatsuji, S., Maeno, Y., Quasi-Two-Dimensional Mott Transition System $Ca_{2-x}Sr_xRuO_4$, *Phys. Rev. Lett.* **84**, 2666 (2000)

21. Mizokawa, T., Tjeng, L.H., Sawatzky, G.A., Ghiringhelli, G., Tjernberg, O., Brookes, N.B., Fukazawa, H., Nakatsuji, S., Maeno, Y., Spin-Orbit Coupling in the Mott Insulator Ca_2RuO_4, *Phys. Rev. Lett.* **87**, 077202 (2001)

22. Snow, C.S., Cooper, S.L., Cao, G., Crow, J.E., Fukazawa, H., Nakatsuji, S., Maeno, Y., Pressure-Tuned Collapse of the Mott-Like State in $Ca_{n+1}Ru_nO_{3n+1}$ (n=1,2): Raman Spectroscopic Studies, *Phys. Rev. Lett.* **89**, 226401 (2002)

23. Qi, T.F., Korneta, O.B., Parkin, S., DeLong, L.E., Schlottmann, P., Cao, G., Negative Volume Thermal Expansion Via Orbital and Magnetic Orders in $Ca_2Ru_{1-x}Cr_xO_4$ (0 < x < 0.13), *Phys. Rev. Lett.* **105**, 177203 (2010)

24. Qi, T.F., Korneta, O.B., Parkin, S., Hu, J., Cao, G., Magnetic and orbital orders coupled to negative thermal expansion in Mott insulators $Ca_2Ru_{1-x}M_xO_4$ (M = Mn and Fe), *Phys. Rev. B.* **85**, 165143 (2012)

25. Moore, R. G., M.D. Lumsden, M.D., M.B. Stone, M.B., Zhang, J.D., Chen, Y., Lynn, J.W., Jin, R. Mandrus, D., Plummer, E. W., Phonon softening and anomalous mode near the x_c=0.5 quantum critical point in $Ca_{2-x}Sr_xRuO_4$, *Phys. Rev. B* **79**, 172301 (2009)

26. Chi, S., Ye, F., Cao, G., Cao, H., Fernandez-Baca, J.A., Competition of three-dimensional magnetic phases in $Ca_2Ru_{1-x}Fe_xO_4$: A structural perspective, *Phys. Rev. B.* **102**, 014452 (2020)

27. Mary, T.A., Evans, J.S.O., Vogt, T., Sleight, A.W., Negative Thermal Expansion from 0.3 to 1050 Kelvin in ZrW_2O_8, *Science.* **272**, 90 (1996)

28. Salvador, J.R., Guo, F., Hogan, T., Kanatzidis, M.G., Zero thermal expansion in YbGaGe due to an electronic valence transition, *Nature.* **425**, 702 (2003)

29. Lakes, R., Cellular solids with tunable positive or negative thermal expansion of unbounded magnitude, *Appl. Phys. Lett.* **90**, 221905 (2007)

30. Goodwin, A.L., Calleja, M., Conterio, M.J., Dove, M.T., Evans, J.S.O., Keen, D.A., Peters, L., Tucker, M.G., Colossal Positive and Negative Thermal Expansion in the Framework Material $Ag_3[Co(CN)_6]$, *Science.* **319**, 794 (2008)

31. Evans, J.S.O., Mary, T.A., Sleight, A.W., Negative Thermal Expansion in a Large Molybdate and Tungstate Family, *Journal of Solid State Chemistry.* **133**, 580 (1997)

32. Fletcher, N.H., *The Chemical Physics of Ice.* (Cambridge University Press, London, 1970)

33. Hemberger, J., Krug von Nidda, H.-A., Tsurkan, V., Loid, A., Large Magnetostriction and Negative Thermal Expansion in the Frustrated Antiferromagnet $ZnCr_2Se_4$, *Phys. Rev. Lett.* **98**, 147203 (2007)

34. Azuma, M., Chen, W.T., Seki, H., Czapski, M., Olga, S., Oka, K., Mizumaki, M., Watanuki, T., Ishimatsu, N., Kawamura, N., Ishiwata, S., Tucker, M.G., Shimakawa, Y., Attfield, J.P., Colossal negative thermal expansion in $BiNiO_3$ induced by intermetallic charge transfer, *Nature Communications.* **2**, 347 (2011)

35. Mazzone, D.G., Dzero, M., Abeykoon, A.M., Yamaoka, H., Ishii, H., Hiraoka, N., Rueff, J.-P., Ablett, J.M., Imura, K., Suzuki, H.S., Hancock, J.N., Jarrig, I., Kondo-Induced Giant Isotropic Negative Thermal Expansion, *Phys. Rev. Lett.* **124**, 125701 (2020)

36. Cao, G., McCall, S., Crow, J.E., Guertin, R.P., Observation of a Metallic Antiferromagnetic Phase and Metal to Nonmetal Transition in $Ca_3Ru_2O_7$, *Phys. Rev. Lett.* **78**, 1751 (1997)

37. Cao, G., Balicas, L., Xin, Y., Dagotto, E., Crow, J.E., Nelson, C.S., Agterberg, D.F., Tunneling magnetoresistance and quantum oscillations in bilayered $Ca_3Ru_2O_7$, *Phys. Rev. B.* **67**, 060406(R) (2003)

38. Cao, G., Balicas, L., Song, W.H., Sun, Y.P., Xin, Y., Bondarenko, V.A., Brill, J.W., Parkin, S., Lin, X.N., Competing ground states in triple-layered $Sr_4Ru_3O_{10}$: Verging on itinerant ferromagnetism with critical fluctuations, *Phys. Rev. B.* **68**, 174409 (2003)

39. Cao, G., Lin, X.N., Balicas, L., Chikara, S., Crow, J.E., Schlottmann, P., Orbitally driven behaviour: Mott transition, quantum oscillations and colossal magnetoresistance in bilayered $Ca_3Ru_2O_7$, *New J. Phys.* **6**, 159 (2004)

40. Lin, X.N., Zhou, Z.X., Durairaj, V., Schlottmann, P., Cao, G., Colossal Magnetoresistance by Avoiding a Ferromagnetic State in the Mott System $Ca_3Ru_2O_7$, *Phys. Rev. Lett.* **95**, 017203 (2005)

41. Durairaj, V., Lin, X.N., Zhou, Z.X., Chikara, S, Ehami, E., Schlottmann, P., Cao, G., Observation of oscillatory magnetoresistance periodic in $1/B$ and B in $Ca_3Ru_2O_7$, *Phys. Rev. B.* **73**, 054434 (2006)

42. Cao, G., Abbound, K., McCall, S., Crow, J.E., Guertin, R.P., Spin-charge coupling for dilute La-doped $Ca_3Ru_2O_7$, *Phys. Rev. B.* **62**, 998 (2000)

43. Liu, H.L., Yoon, S., Cooper, S.L., Cao, G., Crow, J.E., Raman-scattering study of the charge and spin dynamics of the layered ruthenium oxide $Ca_3Ru_2O_7$, *Phys. Rev. B.* **60**, R6980(R) (1999)

44. Karpus, J.F., Gupta, R., Barath, H., Cooper, S.L., Cao, G., Field-Induced Orbital and Magnetic Phases in $Ca_3Ru_2O_7$, *Phys. Rev. Lett.* **93**, 167205 (2004)

45. Gupta, R., Kim, M., Barath, H., Cooper, S.L., Cao, G., Field- and Pressure-Induced Phases in $Sr_4Ru_3O_{10}$: A Spectroscopic Investigation, *Phys. Rev. Lett.* **96**, 067004 (2006)

46. McCall, S., Cao, G., Crow, J.E., Impact of magnetic fields on anisotropy in $Ca_3Ru_2O_7$, *Phys. Rev. B.* **67**, 094427 (2003)

47. Rossiter, P. L., *The Electrical Resistivity of Metals and Alloys* (Cambridge University Press, 1991)

48. Pippard, A.B., *Magnetoresistance in Metals* (Cambridge University Press, 1989)

49. Narayanan, A., Watson, M.D., Blake, S.F., Bruyant, N., Drigo, L., Chen, Y.L., Prabhakaran, D., Yan, B., Felser, C., Kong, T., Canfield, P.C., Coldea, A.I., Linear Magnetoresistance Caused by Mobility Fluctuations in n-Doped Cd_3As_2, *Phys. Rev. Lett.* **114**, 117201 (2015)

50. Leahy, I.A., Lin, Y.-P., Siegfried, P.E., Treglia, A.C., Song, J.C.W., Nandkishore, R.M., Lee, M., Nonsaturating large magnetoresistance in semimetals, *PNAS.* **115**, 10570 (2018)

51. Tokura, Y., *Colossal Magnetoresistive Oxides* (Gordon and Breach Science Publishers, 2000)

52. Sondheimer, E.H., The Influence of a Transverse Magnetic Field on the Conductivity of Thin Metallic Films, *Phys. Rev.* **80**, 401 (1950)

53. Yacoby, A., Schuster, R., Heilblum, M., Phase rigidity and $h/2e$ oscillations in a single-ring Aharonov-Bohm experiment, *Phys. Rev. B.* **53**, 9583 (1996)

54. Schlottmann, P., Zvyagin, A.A., Aharonov–Bohm oscillations at finite temperature, *Journal of Applied Physics.* **79**, 5419 (1996)

55. Schlottmann, P., Aharonov - Bohm - Casher oscillations in strongly correlated electron systems at finite temperature, *J. Phys.: Condens. Matter.* **9**, 7369 (1997)

56. Crawford, M.K., Harlow, R.L., Marshall, W., Li, Z., Cao, G., Lindstrom, R.L., Huang, Q., Lynn, J.W., Structure and magnetism of single crystal $Sr_4Ru_3O_{10}$: A ferromagnetic triple-layer ruthenate, *Phys. Rev. B.* **65**, 214412 (2002)

57. Cao, G., McCall, S., Shepard, M., Crow, J.E., Guertin, R.P., Thermal, magnetic, and transport properties of single-crystal $Sr_{1-x}Ca_xRuO_3$ ($0 \le x \le 1.0$), *Phys. Rev. B.* **56**, 321 (1997)

58. Perry, R.S., Galvin, L.M., Grigera, S.A., Capogna, L., Schofield, A.J., Mackenzie, A.P., Chiao, M., Julian, S.R., Ikeda, S.I., Nakatsuji, S., Maeno, Y., Pfleiderer, C., Metamagnetism and Critical Fluctuations in High Quality Single Crystals of the Bilayer Ruthenate $Sr_3Ru_2O_7$, *Phys. Rev. Lett.* **86**, 2661 (2001)

59. Grigera, S.A., Perry, R.S., Schofield, A.J., Chiao, M., Julian, S.R., Lonzarich, G.G., Ikeda, S.I., Maeno, Y., Millis, A.J., Mackenzie, A.P., Magnetic Field-Tuned Quantum Criticality in the Metallic Ruthenate $Sr_3Ru_2O_7$, *Science.* **294**, 329 (2001)

60. Cao, G., Chikara, S., Brill, J.W., Schlottmann, P., Anomalous itinerant magnetism in single-crystal $Sr_4Ru_3O_{10}$: A thermodynamic and transport investigation, *Phys. Rev. B.* **75**, 024429 (2007)

61. Chikara, S., Durairaj, V., Song, W.H., Sun, Y.P., Lin, X.N., Douglass, A., Cao, G., Schlottmann, P., Parkin, S., Borderline magnetism in $Sr_4Ru_3O_{10}$: Impact of La and Ca doping on itinerant ferromagnetism and metamagnetism, *Phys. Rev. B.* **73**, 224420 (2006)

62. Mao, Z.Q., Zhou, M., Hooper, J., Golub, V., O'Connor, C.J., Phase Separation in the Itinerant Metamagnetic Transition of $Sr_4Ru_3O_{10}$, *Phys. Rev. Lett.* **96**, 077205 (2006)

63. Jo, Y.J., Balicas, L., Kikugawa, N., Choi, E.S., Storr, K., Zhou, M., Mao, Z.Q., Orbital-dependent metamagnetic response in $Sr_4Ru_3O_{10}$, *Phys. Rev. B.* **75**, 094413 (2007)

64. Carleschi, E., Doyle, B.P., Fittipaldi, R., Granata, V., Strydom, A.M., Cuoco, M., Vecchione, A., Double metamagnetic transition in $Sr_4Ru_3O_{10}$, *Phys. Rev. B.* **90**, 205120 (2014)

65. Moriya, T., *Spin Fluctuations in Itinerant Electron Magnetism* (Springer-Verlag Berlin Heidelberg, 1985)

66. Fazekas, P., *Lecture Notes On Electron Correlation And Magnetism* (World Scientific, 1999)

67. Stoner, E.C., Collective electron ferronmagnetism, *Proceedings of the Royal Society A. Mathematical, Physical and Engineering Sciences.* **165**, 372 (1938)

68. Wohlfarth, E.P., Rhodes, P., Collective electron metamagnetism, *The Philosophical Magazine: A Journal of Theoretical Experimental and Applied Physics.* **7**, 1817 (1962)

69. Yamada, H., Metamagnetic transition and susceptibility maximum in an itinerant-electron system, *Phys. Rev. B.* **47**, 11211 (1993)

70. Mirri, C., Vitucci, F.M., Di Pietro, P., Lupi, S., Fittipaldi, R., Granata, V., Vecchione, A., Schade, U., Calvani, P, Anisotropic optical conductivity of $Sr_4Ru_3O_{10}$, *Phys. Rev. B.* **85**, 235124 (2012)

71. Malvestuto, M., Capogrosso, V., Carleschi, E., Galli, L., Gorelov, E., Pavarini, E., Fittipaldi, R., Forte, F., Cuoco, M., Vecchione, A., Parmigiani, F., Nature of the apical and planar oxygen bonds in the $Sr_{n+1}Ru_nO_{3n+1}$ family ($n=1,2,3$), *Phys. Rev. B.* **88**, 195143 (2013)

72. Weickert, F., Civale, L., Maiorov, B., Jaime, M., Salamon, M.B., Carleschi, E., Strydom, A.M., Fittipaldi, R., Granata, V., Vecchione, A., Missing magnetism in $Sr_4Ru_3O_{10}$: Indication for Antisymmetric Exchange Interaction, *Scientific Reports.* **7**, 3867 (2017)

73. Granata, V., Capogna, L., Reehuis, M., Fittipaldi, R., Ouladdiaf, B., Pace, S., Cuoco, M., Vecchione, A., Neutron diffraction study of triple-layered $Sr_4Ru_3O_{10}$, *J. Phys.: Condens. Matter.* **25**, 056004 (2013)

74. Liu, Y., Yang, J., Wang, W., Du, H., Ning, W., Ling, L., Tong, W., Qu, Z., Yang, Z., Tian, M., Cao, G., Zhang, Y., Size effect on the magnetic phase in $Sr_4Ru_3O_{10}$, *New J. Phys.* **18**, 053019 (2016)

75. Zheng, H., Song, W.H., Terzic, J., Zhao, H.D., Zhang, Y., Ni, Y.F., DeLong, L.E., Schlottmann, P., Cao, G., Observation of a pressure-induced transition from interlayer ferromagnetism to intralayer antiferromagnetism in $Sr_4Ru_3O_{10}$, *Phys. Rev. B.* **98**, 064418 (2018)

76. Mazin, I.I., Singh, D.J., Electronic structure and magnetism in Ru-based perovskites, *Phys. Rev. B.* **56**, 2556 (1997)

77. Cuoco, M., Forte, F., Noce, C., Probing spin-orbital-lattice correlations in $4d^4$ systems, *Phys. Rev. B.* **73**, 094428 (2006)

78. Cao, G., Durairaj, V., Chikara, S., DeLong, L.E., Schlottmann, P., Observation of Strong Spin Valve Effect in Bulk $Ca_3(Ru_{1-x}Cr_x)_2O_7$, *Phys. Rev. Lett.* **100**, 016604 (2008)
79. Shlyk, L., Parkin, S., DeLong, L.E., Successive magnetic transitions of the kagome plane and field-driven chirality in single-crystal $BaMn_{2.49}Ru_{3.51}O_{11}$, *Phys. Rev. B.* **81**, 014413 (2010)
80. Hamlin, J.J., Deemyad, S., Schilling, J.S., Jacobsen, M.K., Kumar, R.S., Cornelius, A.L., Cao, G., Neumeier, J.J., ac susceptibility studies of the weak itinerant ferromagnet $SrRuO_3$ under high pressure to 34 GPa, *Phys. Rev. B.* **76**, 014432 (2007)
81. Liu, Y., Chu, W., Yang, J., Liu, G., Du, H., Ning, W., Ling, L., Tong, W., Qu, Z., Cao, G., Xu, Z., Tian, M., Magnetic reversal in $Sr_4Ru_3O_{10}$ nanosheets probed by anisotropic magnetoresistance, *Phys. Rev. B.* **98**, 024425 (2018)

Chapter 5

Current Control of Structural and Physical Properties in Spin-Orbit-Coupled Mott Insulators

5.1 Overview

Experimental condensed matter and materials research typically focuses on the responses of electrons and/or phonons to applied stimuli; studies of quantum materials under extreme conditions of high magnetic field, high pressure, or ultralow temperatures are good examples [1,2]. In $4d$- and $5d$-transition metal oxides, a delicate interplay between fundamental interactions (see Table 1.2) leaves these materials extremely susceptible to small external stimuli, as schematically illustrated in **Fig. 5.1**. In particular, these stimuli can strongly couple to the lattice, resulting in novel quantum states [1–6], some of which are discussed in previous chapters.

Electrical current as a means to control structural and related physical properties has been recognized only recently. The application of small electrical currents in sensitive detector and control applications, and in information technologies is often preferable to other external stimuli in part because it is more accessible. However, it has not been widely accepted that electrical current can readily couple to the lattice, orbital, or spin degrees of freedom until recently. Nevertheless, mounting experimental evidence has indicated that a combination of strong spin-orbit interactions (SOI) and a distorted crystal structure in magnetic Mott insulators may be sufficient for electrical current to control structural and related properties, as schematically shown in **Fig. 5.2**. Current control of quantum states in $4d$- and $5d$-transition metal oxides has therefore rapidly expanded as a key research topic [7–13].

In fact, electrical current is surprisingly effective in coupling with the lattice, as observed in the physical properties of certain $4d$- and $5d$-transition metal oxides with magnetic canting. In essence, this is because strong SOI lock magnetic moments to the lattice via the extended/directed d-orbitals, which facilitate a strong coupling of current with electron orbitals. Empirical evidence indicates that a small applied electrical current (e.g., current density ~ 0.15 A/cm^2) can critically reduce structural distortions, which in turn readily suppresses the antiferromagnetic (AFM) insulating state, and, in some

Physics of Spin-Orbit-Coupled Oxides. Gang Cao and Lance E. DeLong, Oxford University Press (2021). © Gang Cao and Lance E. DeLong.
DOI: 10.1093/oso/9780199602025.003.0005

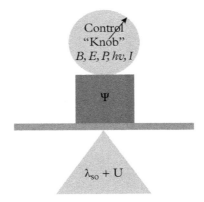

Fig. 5.1 *Schematic illustration of the high susceptibility of a quantum state Ψ supported by λ_{so} and U to external stimuli, such as magnetic field B, electrical field E, pressure P, light hv, or electrical current I.*

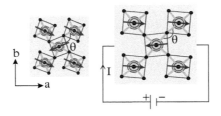

Fig. 5.2 *Schematic illustration of current-controlled structural (e.g., octahedral tilt angle θ) and magnetic properties. Left: Arrows are magnetic moments strongly locked with the lattice. Right: Applied current I engages with the lattice to relax lattice distortions via SOI [9].*

cases, results in new emergent states. In other words, current-reduced lattice distortions dictate physical properties via SOI; an effective current control of quantum states is not anticipated in materials with low atomic number Z. For example, the SOI ($\sim Z^2$) is much weaker in $3d$-transition metal oxides (note that $3d$-orbitals are more contracted compared to those in their $4d$- and $5d$-counterparts; see **Fig. 1.1** and Table 1.2).

Early studies already suggested that electronic properties of certain $4d$- and $5d$-transition metal oxides are sensitive to applied electrical current. In the late 1990s, it was found that $Ca_3Ru_2O_7$ (**Fig. 5.3**) [14], $BaIrO_3$ (**Fig. 5.4**) [15], and Sr_2IrO_4 (**Fig. 2.8**) [16] exhibit an S-shaped negative differential resistivity (NDR). NDR is generally attributed to either an "electro-thermal effect" or a "transferred carrier effect" in which a current drives carriers from a high- to a low-mobility band, as in the Gunn effect. The more common form of NDR is manifest in N-shaped characteristics [17–23]. Alternatively, an S-shaped NDR has been observed in memory devices and a few bulk materials such as VO_2,

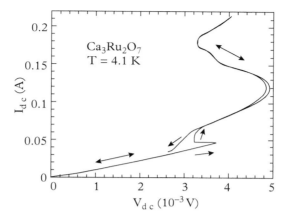

Fig. 5.3 *$Ca_3Ru_2O_7$: Current-voltage characteristics: S-shaped NDR [14].*

$CuIr_2S_{4-x}Se_x$, and 1T-TaS_2 [17–23]. All of these bulk materials are characterized by a first-order metal-insulator transition without an AFM state. It is therefore peculiar for AFM Sr_2IrO_4 and $BaIrO_3$ to show the S-shaped NDR because these materials show no first-order metal-insulator transition. Furthermore, an early study also revealed a current-induced metallic state in Sr_2IrO_4 (**Fig. 2.8**, inset). A current-reduced resistivity in Ca_2RuO_4 was also reported in a more recent study [24]. These studies all signaled that a combined effect of SOI and Coulomb interactions in 4d- and 5d-transition oxides may result in an unusually high susceptibility to application of electrical current, which has helped motivate more extensive studies on this topic in recent years.

Application of a small electrical current differs from application of static or low-frequency electrical fields, which yield electrical-field-controlled magnetic properties in multiferroics. This important but vastly different topic has been extensively studied over the last two decades [25,26, for example].

Alternatively, the Poole-Frenkel effect, which successfully describes electrical-field-induced conductivity in many semiconductors and insulators [27], is unlikely to apply to the current-induced phenomena in spin-orbit-coupled Mott insulators discussed here: the Poole-Frenkel effect is a result of ionization by an applied electrical field that increases the number of conduction electrons and, therefore, the conductivity of the material [27]. In contrast, application of small electrical current leads to drastic anisotropic changes in the structural and magnetic properties of spin-orbit-coupled Mott insulators.

It is always important to check for effects due to Joule heating, which could cause spurious anomalies in electric transport behavior. Therefore, studies of current-controlled phenomena require robust, innovative techniques that eliminate effects of Joule heating.

In the following sections, current-controlled phenomena observed in two model materials, Ca_2RuO_4 and Sr_2IrO_4, are discussed.

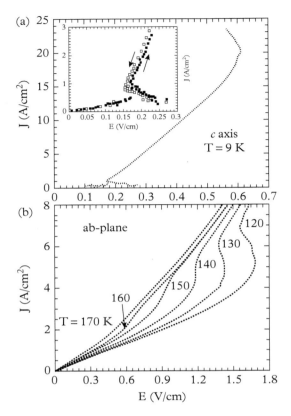

Fig. 5.4 *BaIrO$_3$: S-shaped NDR for current along (a) the c-axis and (b) the ab-plane. The inset shows details of the I-V characteristics at low current and the Ohmic behavior for I < 2 mA [15].*

5.2 Ca$_2$RuO$_4$

As discussed in Chapter 4, the lattice-driven Ca$_2$RuO$_4$ exhibits a metal-insulator transition at T$_{MI}$ = 357 K driven by a violent structural transition [28] characterized by a severe rotation and tilting of RuO$_6$ octahedra, which results in considerably enhanced orthorhombicity below T$_{MI}$, and lifting the t_{2g} orbital (d_{xy}, d_{yz}, d_{zx}) degeneracy. An AFM state is only stabilized at a much lower temperature, T$_N$ = 110 K, by a further rotation and tilting of the RuO$_6$ octahedra. It is now well recognized that physical properties are dictated by the structural distortions and the populations of t_{2g} orbitals, particularly, the d_{xy}-orbitals [30–42].

In this investigation, Ca_2RuO_4 with 3% Mn doping $(Ca_2Ru_{0.97}Mn_{0.03}O_4)$ is chosen because the dilute Mn doping weakens the violent first-order structural phase transition at 357 K, but conveniently preserves the underlying structural and physical properties of Ca_2RuO_4. The doped single crystals are therefore more robust against the thermal cycling necessary for thorough electric transport measurements [43,44].

The key element in this study is a strong coupling between a small applied electrical current and the lattice that reduces the orthorhombic distortion, the octahedral rotation, and the tilt, which are crucial to the population of the t_{2g} orbitals. These structural changes readily suppress the native AFM insulating state and subsequently induce a nonequilibrium orbital state. A phase diagram based on the data reveals a narrow critical regime near a small current density of 0.15 A/cm² that separates the native, diminishing AFM state and the emergent, nonequilibrium orbital state [8].

5.2.1 Current-Controlled Physical Properties

The magnetization, M, of $Ca_2Ru_{0.97}Mn_{0.03}O_4$ is highly sensitive to a small current density, J. As seen in **Figs. 5.5a** and **5.5b**, the *a*- and *b*-axis magnetizations, M_a and M_b, change drastically with electrical current applied along the *b* axis. The AFM transition T_N drops rapidly from 125 K at current density J = 0 A/cm² to 29 K at J = 0.15 A/cm² in M_b, and completely disappears at J > J_C ~ 0.15 A/cm² [8]. (Note that 0.15 A/cm² is a remarkably small current density!) The suppression of the AFM state is accompanied by a drastic decrease in the *b*-axis resistivity, ρ_b, by up to four orders of magnitude, as shown in **Fig. 5.5c**.

Upon the vanishing of the AFM transition T_N, a distinct phase emerges below $T = T_O \approx 80$ K (**Fig. 5.6a**). The phase transition temperature T_O initially rises slightly, peaks at J = 0.28 A/cm² (**Fig. 5.6a**) before decreasing with further increases of J. The *b*-axis resistivity ρ_b closely tracks the magnetization M_a, as shown in **Fig. 5.6b**. The simultaneous changes in both ρ_b and M_a at T_O indicates a strong correlation between the transport and magnetic properties in the current-induced state, which sharply contrasts with the native state in which the onset of AFM order T_N is nearly 250 K below T_{MI}, implying that the nature of the current-induced state is distinctly different from that of the native state. Note that the current-induced state emerges only when the native AFM state completely vanishes.

5.2.2 Current-Controlled Structural Properties

Application of electrical current to the system effectively relaxes structural distortions. With increasing J, the orthorhombicity, defined by $(b-a)/[(a+b)/2]$, readily decreases: the temperature dependence of the lattice parameters *a* and *b* at J = 0 and 4 A/cm² clearly show a diminishing orthorhombicity with increasing J (see **Fig. 5.7**). More detailed structural data indicate that the difference between the *a* and *b* parameters decreases with increasing J (**Fig. 5.8a**); thus, the orthorhombicity weakens from 4.4% at J = 0 A/cm² to 1.2% at J = 30 A/cm² (**Fig. 5.8b**). Moreover, the *c* axis expands by up to 2.4% at J = 30 A/cm²

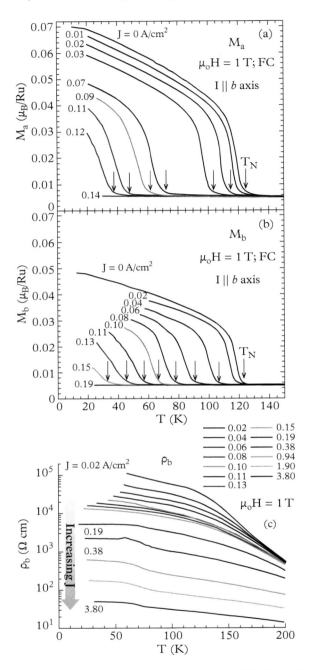

Fig. 5.5 *Current-driven magnetic and electric transport properties of $Ca_2Ru_{0.97}Mn_{0.03}O_4$: The temperature dependence at various current densities J applied along the b axis of (a) the a-axis magnetization M_a, (b) the b-axis magnetization M_b, and (c) the b-axis resistivity ρ_b. The magnetic field $\mu_oH = 1$ T [8].*

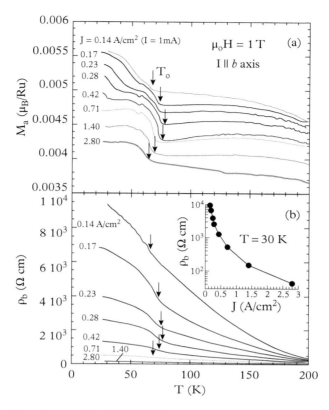

Fig. 5.6 *Current-induced state at J ≥ 0.14 A/cm²: the temperature dependence at various J applied along the b axis of (a) M_a at 1 Tesla and (b) ρ_b for $Ca_2Ru_{0.97}Mn_{0.03}O_4$. Inset: ρ_b at 30 K as a function of J [8].*

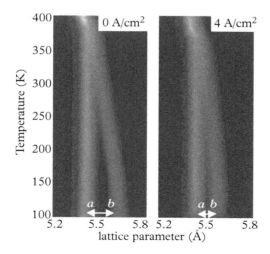

Fig. 5.7 *The current-reduced orthorhombicity in $Ca_2Ru_{0.97}Mn_{0.03}O_4$: Two representative contour plots for the temperature dependence of the lattice parameters a and b at current density J = 0 and 4 A/cm² applied in the basal plane. Note the diminishing orthorhombicity with increasing J marked by the white arrows [8].*

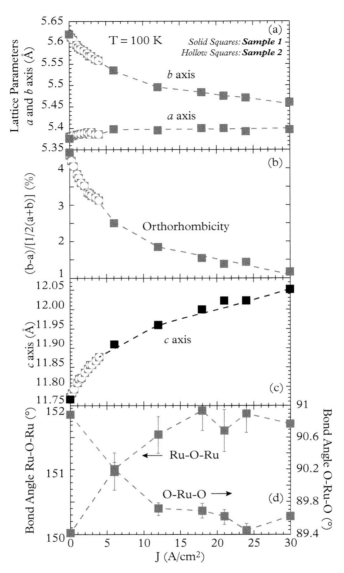

Fig. 5.8 *Neutron diffraction for current-driven lattice changes in $Ca_2Ru_{0.97}Mn_{0.03}O_4$: the current density J dependence at 100 K of (a) the a- and b-axis, (b) the orthorhombicity, (c) the c-axis, and (d) the bond angles Ru-O-Ru (left scale) and O-Ru-O (right scale) [8].*

(**Fig. 5.8c**). It is also crucial that the bond angle Ru-O-Ru, which defines the rotation of RuO_6 octahedra, increases by up to two degrees at $J = 18$ A/cm², giving rise to a much-reduced distortion (**Fig. 5.8d**). Finally, the bond angle O-Ru-O decreases from 91° to 90.2° at $J = 5$ A/cm², which is close to the ideal O-Ru-O bond angle of 90° (**Fig. 5.8d**).

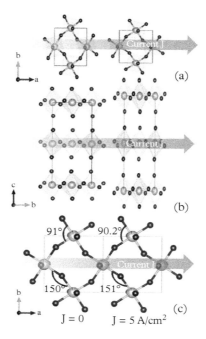

Fig. 5.9 *Schematic illustrations of the current-induced lattice changes: (a) reduced orthorhombicity, (b) elongated c axis, and (c) relaxed bond angles [8].*

Figure 5.9 summarizes the effect of applied current on the lattice: the applied current J weakens the orthorhombicity (**Fig. 5.9a**), elongates the *c*-axis (**Fig. 5.9b**), and relaxes the bond angles toward a less-distorted structure (**Fig. 5.9c**).

5.2.3 Correlations between Current-Controlled Structural and Physical Properties

The simultaneous measurements of neutron diffraction and electrical resistivity document the direct link between the current-reduced orthorhombicity and resistivity (**Fig. 5.10**). The orthorhombicity as functions of temperature and current density ranging from 0 to 4 A/cm² in **Fig. 5.10a** shows that the orthorhombic distortion rapidly declines with current density J. The resistivity almost perfectly tracks the orthorhombicity, as shown in **Fig. 5.10b**: the contour-plot comparison of **Figs. 5.10a** and **5.10b** establishes an explicit correlation between the current-driven lattice and transport properties.

Remarkably, the structural transition T_{MI}, which is defined by the blue area in **Fig. 5.10a**, hardly shifts with J. The data further confirm that Joule heating effect is inconsequential in these studies [7–9].

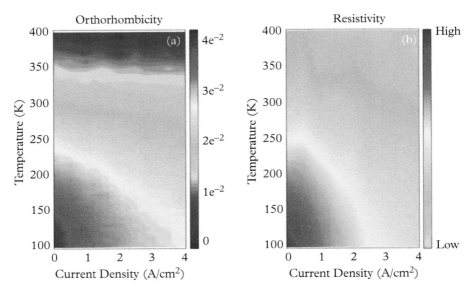

Fig. 5.10 *Direct correlation between the orthorhombicity and the electrical resistivity of $Ca_2Ru_{0.97}Mn_{0.03}O_4$: The temperature-current-density contour plots for (a) the orthorhombicity and (b) electrical resistivity [8].*

5.2.4 Current-Induced Nonequilibrium Orbital State

A temperature versus current-density phase diagram based on the data presented earlier summarizes the current-controlled phenomena in Ca_2RuO_4: by reducing the structural distortions and changing the t_{2g} orbital occupancies, the applied current effectively destabilizes the insulating and AFM state and, over a narrow critical regime of current density J_C, precipitates the nonequilibrium orbital state, as shown in **Fig. 5.11**.

In ambient conditions, the tetravalent Ru^{4+} ion with four $4d$ electrons provides two holes in the t_{2g} orbitals (with empty e_g orbitals). One-half hole is transferred to the oxygen [8], and the remaining $3/2$ hole is equally split in a 1:1 ratio between the d_{xy}-orbital and the manifold of d_{xz}/d_{yz} orbitals at high temperatures or in the metallic state at $T > T_{MI}$. At $T < T_{MI}$, the first-order transition $T_{MI} = 357$ K enhances the orthorhombicity and other distortions including the rotation, tilting, and flattening of RuO_6 octahedra. These changes facilitate a transfer of more holes from d_{xy} to d_{xz}/d_{yz}, or a 1:2 ratio of hole occupancies in d_{xy} vs. d_{xz}/d_{yz} [3,8]. The insulating state below T_{MI} thus has each orbital at exactly $3/4$ electron filling. This contrasts with the metallic state above T_{MI} which has a nearly half-filled d_{xy}-orbital and unequal filling, with a nearly filled d_{xz}/d_{yz} manifold.

The half-filled d_{xy}-orbital is the key to the metallic state. The applied current helps stabilize the existence of the half-filled d_{xy}-orbital as temperature decreases by minimizing structural distortions. These current-induced lattice changes also explain the suppression of the native AFM state with increasing J because it delicately depends on a

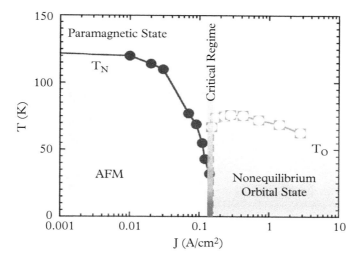

Fig. 5.11 *The T-J phase diagram illustrates that the applied current drives the system from the native AFM state through the critical regime near 0.15 A/cm² to the current-induced nonequilibrium orbital state [8].*

combination of rotation, tilt, and flattening of RuO_6 octahedra, all of which are significantly weakened by applied current.

While the understanding of the nonequilibrium orbital state at $J > J_c$ (**Fig. 5.11**) is yet to be fully established, it is clear that the critical lattice modifications are at the heart of the current-controlled phenomena via current-driven, nonequilibrium orbital populations.

5.3 Sr₂IrO₄

The underlying properties of Sr_2IrO_4 are discussed in Chapter 2. This spin-orbit-coupled Mott insulator has an AFM transition at $T_N = 240$ K and an electronic energy gap $\Delta < 0.62$ eV [45, 46]. It adopts a tetragonal structure with $a = b = 5.4846$ Å and $c = 25.804$ Å with space group $I4_1/a$ (No. 88) [47–49].

There are two signature characteristics essential to the current-controlled behavior: (1) Rotation of the IrO_6 octahedra about the c axis by approximately 12°, which corresponds to a distorted in-plane Ir1-O2-Ir1 bond angle θ and has a critical effect on the ground state. (2) The magnetic structure features canted moments (0.208(3) μ_B/Ir) within the basal plane [47]. This 13(1)°-canting of the moments away from the a axis closely tracks the staggered rotation of the IrO_6 octahedra [48, 49] (see **Fig. 2.5**), suggesting a strong interlocking of the magnetic moments with the lattice.

The relationship between the IrO_6 rotation and magnetic canting in the iridate was first discussed in [50], in which a theoretical model proposed a strong magnetoelastic coupling in Sr_2IrO_4 and a close association between the magnetic canting and the c/a ratio, as a result of the strong SOI. Such a strong locking of the moment canting to the IrO_6-rotation (by $11.8(1)°$) was confirmed in later studies of neutron diffraction [47], X-ray resonant scattering [48], and second-harmonic generation (SHG) [49]. SHG or frequency doubling is an optical process where two photons with the same frequency interact with a material and create a new photon with twice the frequency of the initial photons (see Appendix F).

As demonstrated in the following, the strong interlocking between the lattice and magnetic moments is the key for the current control of quantum states in Sr_2IrO_4.

5.3.1 Current-Controlled Structural Properties

The study of current-controlled structural properties of Sr_2IrO_4 [7] was stimulated by early experimental observations of the S-shaped I-V characteristic and reduced resistivity (see Chapter 2), and, more recently, by observation of a current-induced color change of single-crystal Sr_2IrO_4 under illumination by polarized light, which eventually led to comprehensive studies of the properties of this iridate as functions of both electrical current and temperature [7].

The structural response to applied current, I, is shown in **Fig. 5.12**, as reflected in changes of representative Bragg peaks at T = 200 K with applied current within the basal

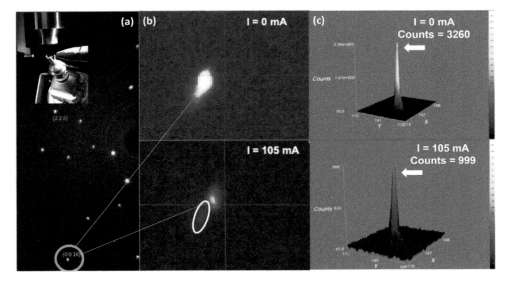

Fig. 5.12 Sr_2IrO_4: *Single-crystal X-ray diffraction with current I applied within the basal plane of the crystal: (a) representative X-ray diffraction pattern of single-crystal Sr_2IrO_4. The circled Bragg peak is (0016). Inset: Sample mounting showing electrical leads and cryogenic gas feed. (b) Contrasting the (0016) peak location for I = 0 mA (upper panel) and I = 105 mA (lower panel). The white oval outline marks the peak location for I = 0 mA for comparison. (c) The intensity of the (0016) peak is 3260 for I = 0 mA (upper panel) and 999 for I = 105 mA (lower panel) [7].*

plane. A close examination of (0016)-peak (**Fig. 5.12a**) reveals shifts in both position and intensity at an applied current I = 105 mA (see **Figs. 5.12b** and **5.12c**). This peak undergoes a threefold reduction in intensity from 3260 counts at I = 0 mA to 999 counts at I = 105 mA (see **Fig. 5.12c**), which suggests that significant shifts have occurred in the atomic positions. Similar changes are seen in other Bragg peaks [7].

More detailed investigations of the crystal structure as functions of both current and temperature reveal an unexpectedly large lattice expansion due to applied current. In particular, at I = 105 mA the a axis elongates by nearly 1% ($\Delta a/a \equiv [a(I)-a(0)]/a(0) = 1\%$) near T_N = 240 K (see **Fig. 5.13a**). In contrast, the c axis changes only very slightly ($\Delta c/c$ < 0.1%) at the same current. The contrasting response of the a and c axes to current I indicates an important role for the basal-plane magnetic moments. Indeed, the temperature

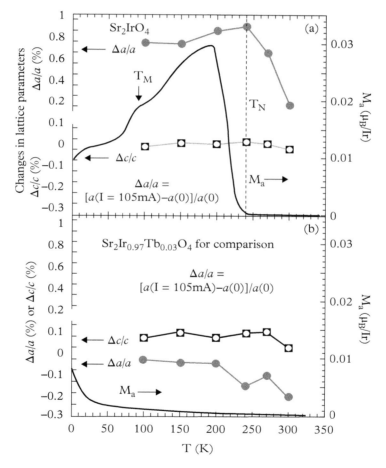

Fig. 5.13 Sr_2IrO_4: *Current-controlled changes $\Delta a/a$ and $\Delta c/c$ and the a-axis magnetization M_a (right scale) for (a) AFM Sr_2IrO_4 and (b) isostructural paramagnetic $Sr_2Ir_{0.97}Tb_{0.03}O_4$ for comparison to (a). Note that the scales for $\Delta a/a$, $\Delta c/c$, and M_a are the same in (a) and (b) for contrast and comparison [7].*

dependence of $\Delta a/a$ closely tracks that of the a-axis magnetization, M_a whereas $\Delta c/c$ is essentially temperature-independent (**Fig. 5.13a**).

This is further confirmed by a controlled study of isostructural, paramagnetic $Sr_2Ir_{0.97}Tb_{0.03}O_4$, in which a 3% substitution of Tb^{4+} for Ir^{4+} leads to a disappearance of T_N but conveniently preserves the original crystal structure and the insulating state [51] (3% Tb doping completely suppresses the AFM, which reaffirms the high sensitivity of the magnetic properties to slight lattice changes). An energy level mismatch between the Ir and Tb sites weakens charge carrier hopping and causes a persistent insulating state [51]. This study indicates that changes in the lattice parameters or absolute values of $\Delta a/a$ and $\Delta c/c$ at I = 105 mA are very small (< 0.2%) and essentially temperature- independent (**Fig. 5.13b**). The sharp contrast between **Figs. 5.13a** and **5.13b** clearly points out a crucial role for the long-range AFM in the current-induced lattice expansion [7]. Without application of current, the a axis expands by no more than ~ 0.1% from 90 K to 300 K, due to conventional thermal expansion, in contrast to the 1% increase due to application of current.

The observations reveal that lattice parameters respond differently to current I applied to the a axis and c axis. As shown in **Fig. 5.14**, the b axis expands up to 0.8% near T_N for I applied along the a axis (**Fig. 5.14a**) but only half that value for I applied along the

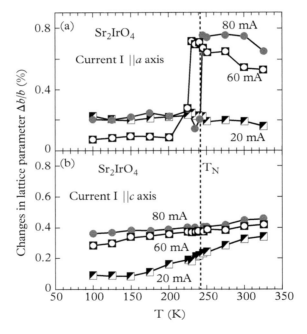

Fig. 5.14 Sr_2IrO_4: *Anisotropic response of the b-axis lattice parameter to current I applied along (a) the a axis and (b) the c axis. Note that the scale for $\Delta b/b$ is the same for both (a) and (b) to facilitate comparisons. The effect of current applied along the a axis is twice as strong as that of current applied along the c axis. The abrupt expansion of the b axis near T_N observed in (a) further underscores the current effect and eliminates any role for Joule heating [9].*

c-axis (**Fig. 5.14b**). Moreover, the abrupt jump in $\Delta b/b$ near T_N tracks the magnetization $M(T)$, further underscoring the interlocking of the canted moments to the lattice when I is along the *a* axis (**Fig. 5.14a**). In contrast, $\Delta b/b$ for I along the *c* axis does not exhibit such anomalous behavior, which implies a much weaker coupling of current and magnetization in this geometry (**Fig. 5.14b**).

It is important to note the anisotropic response also rules out any effect of Joule heating. Should Joule heating play an important role, then its effect would be more or less uniform or isotropic, rather than anisotropic, as seen in **Fig. 5.14**; additionally, the heating effect would be significantly stronger when current is applied along the *c*-axis because the *c*-axis resistivity is at least two orders of magnitude greater than the *a*-axis resistivity.

5.3.2 Current-Controlled Magnetic Properties

Because of the strong coupling between the lattice and magnetic moments [47–50], it is clear that the current-induced lattice expansion must cause changes in magnetic properties. As shown in **Fig. 5.15**, both the *a*-axis magnetic susceptibility $\chi_a(T)$ and the *a*-axis magnetization M_a strongly respond to the current applied along the *a*-axis: the AFM transition T_N is suppressed by 40 K at I = 80 mA (**Figs. 5.15a** and **5.15b**) and the isothermal magnetization M_a is reduced by 16% (**Fig. 5.15c**). Magnetic canting is described by the Dzyaloshinsky-Moriya interaction, i.e., $\mathbf{D}\cdot(\mathbf{S_i}\times\mathbf{S_j})$; the vector $\mathbf{D}$, which measures distortions, approaches zero when the neighboring spins $\mathbf{S_i}$ and $\mathbf{S_j}$ become collinearly

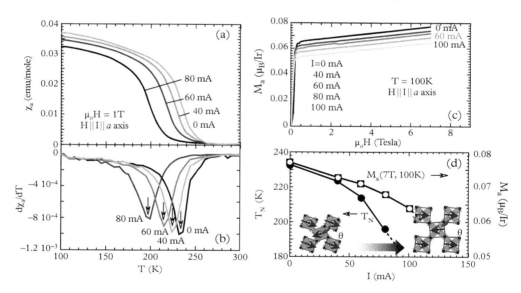

Fig. 5.15 *Sr₂IrO₄: Temperature dependence of (a) a-axis magnetic susceptibility $\chi_a(T)$ at a few representative currents, and (b) $d\chi_a(T)/dT$ clarifying the decrease in T_N with I. (c) $M_a(H)$ at 100 K for a few representative currents. (d) Current dependence of T_N and M_a. Diagrams schematically illustrate the current-controlled lattice expansion, θ and Ir moments (arrows) with I [7].*

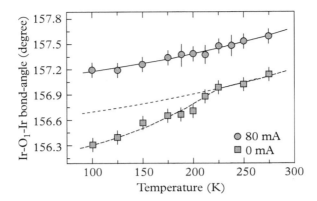

Fig. 5.16 *Sr_2IrO_4: Temperature dependence of the basal-plane Ir-O-Ir bond angle with the applied current I = 0 mA (squares) and 80 mA (circles), respectively [52].*

aligned (also see Section 1.7). The magnetic changes presented in **Fig. 5.15** are therefore understandable because the applied current relaxes the Ir-O-Ir bond angle θ, which weakens magnetic canting and the overall AFM state, as schematically illustrated in **Fig. 5.15d**. Indeed, a neutron diffraction study on Sr_2IrO_4 confirms that application of electrical current significantly increases the Ir-O-Ir bond angle (**Fig. 5.16**) [52].

5.3.3 Non-Ohmic I-V Characteristics

The current-induced lattice expansion also enhances electron mobility in general, and precipitates an unusual quantum switching effect in particular. As shown in **Figs. 5.17a–5.17c**, a linear I-V response during an initial current increase is followed by a sharp threshold voltage V_{th}, which marks a switching point where V abruptly drops with increasing I. This switching point is followed by another broad turning point that emerges at a higher current. A strong anisotropy in the I-V characteristics for the *a*-axis (**Fig. 5.17a**) and the *c*-axis (**Fig. 5.17b**) is illustrated in **Fig. 5.17c** [7].

Interestingly, the threshold voltage V_{th} as a function of temperature shows a distinct slope change near a magnetic anomaly $T_M \approx 100$ K [53] (see **Fig. 5.18**). Early studies have demonstrated that the magnetization M_a undergoes additional anomalies at $T_M \approx$ 100 K and 25 K (**Fig. 5.18**, right scale), due chiefly to moment reorientations [53]. This magnetic reorientation separates the different regimes of I-V behavior below and above $T_M \approx 100$ K (**Fig. 5.18**). The concurrent changes in both V_{th} and M_a strongly indicate a close correlation between the I-V characteristics and magnetic state, and, more generally, a mechanism that is fundamentally different from that operating in other materials [18].

Nevertheless, the current-controlled *a*-axis expansion $\Delta a/a$ (upper horizontal axis in **Fig. 5.19a**) closely tracks the I-V curves with nonlinear changes at two critical currents I_{C1} (= 10 mA) and I_{C2} (= 45 mA), respectively [7]. The slope changes in $\Delta a/a$ indicate successively more rapid expansions of the *a* axis at I_{C1} and I_{C2}, and each of them is accompanied by a more significant increase in the Ir-O-Ir bond angle θ, which in turn improves electron hopping (**Fig. 5.19b**). The *a*-axis expansion $\Delta a/a$ seems to saturate as

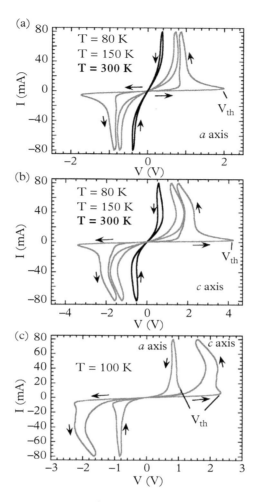

Fig. 5.17 *Sr₂IrO₄: I-V curves at representative temperatures for (a) current applied along the a axis, (b) the c axis, and (c) both the a and c axes at T = 100 K. Arrows show the evolution of the current sweeps in (a) to (c) [7].*

the current further increases above I_{C2} = 45 mA, suggesting that the lattice parameters cannot further expand at $I > I_{C2}$. This explains why a magnetic field H reduces V considerably only between I_{C1} and I_{C2} but shows no visible effect above I_{C2} (**Fig. 5.19a**), where the saturation of $\Delta a/a$ corresponds to θ approaching 180°, which prevents further increases.

The correlations between the *a*-axis expansion, the magnetic canting, and the I-V curves highlights a crucial role of the current-controlled basal-plane expansion that dictates the quantum states. In essence, this is because the IrO_6-octahedra and the canted moment are locked together, thus rigidly rotate together due to strong SOI; applied

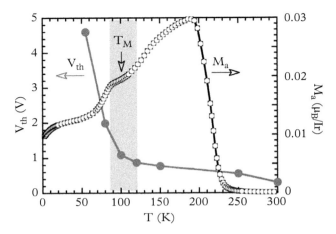

Fig. 5.18 *Sr_2IrO_4: Correlation between the switching effect and magnetization: temperature dependence of the threshold voltage V_{th} for the a-axis and a-axis magnetization $M_a(T)$. Note the slope change of V_{th} near T_M [9].*

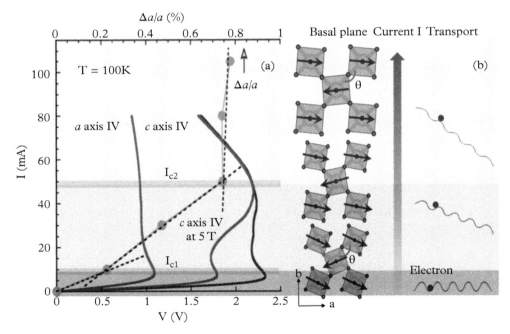

Fig. 5.19 *Sr_2IrO_4: (a) I-V curves for a and c axes at T = 100 K and applied magnetic field of 0 or 5 Tesla along the c axis. The upper horizontal axis is current-controlled a-axis expansion $\Delta a/a$ at T = 100 K. Dashed lines are guides to the eye. Note slope changes of $\Delta a/a$ occur at the two turning points of the I-V curves at I_{C1} and I_{C2}, respectively. (b) Diagrams (not to scale) illustrate the expanding lattice, increasing Ir1-O2-Ir1 bond angle θ and decreasing magnetic canting (arrows) with increasing I. The reduced lattice distortions lead to enhanced electron mobility (see schematic) [7].*

electrical current effectively engages with the lattice and expands the basal plane by increasing θ, which in turn reduces the magnetic canting and the AFM transition T_N and enhances the electron mobility, as illustrated in **Fig. 5.19**.

5.4 General Remarks

It has become increasingly clear that current-controlled phenomena are widespread and present in a range of high-Z materials including both oxides and chalcogenides having an (anti)ferromagnetic and insulating state.

A few empirical trends are worth noting.

First, current-controlled materials must have comparable SOI and U and a ground state that is both (anti)ferromagnetic and insulating. It is apparent that 4d- and 5d-materials provide an energy setting more desirable for current control of quantum states because the electronic wave functions are more extended, the d-band width is wider, and U and the Hund's rule coupling J_H are smaller than those in the 3d-transition metal oxides (Table 1.2). In these materials, the insulating and magnetic ground state are not enabled by large U but rather driven by subtle interactions that are assisted by SOI; thus, a small external stimulus such as electrical current could be sufficient to produce a large response, leading to phase transitions.

Second, an effective current control of quantum states also requires the presence of a distorted crystal structure, often times, canted moments and a strong magnetoelastic coupling. A distorted structure such as rotations and/or tilts of octahedra MO_6 generates room for applied current to relax correlated octahedral rotations and/or tilts and magnetic structures via a strong magnetoelastic coupling that locks magnetic moments with the lattice through SOI. This point is demonstrated in $Sr_3Ir_2O_7$ [54] in which current-controlled behavior is essentially absent. This double-layered iridate is a sister compound of Sr_2IrO_4 and an AFM insulator with $T_N = 285$ K and equally strong SOI. However, the magnetic moments in $Sr_3Ir_2O_7$ are collinearly aligned along the c-axis [55,56] rather than canted within the basal plane where the octahedral rotation occurs, contrasting with that of Sr_2IrO_4 (see Chapter 2). As a result, applied current cannot exert a sufficient effect because the collinearly aligned magnetic moments along the c axis are not strongly coupled with the IrO_6 rotation. Therefore, $Sr_3Ir_2O_7$ hardly shows current-controlled behavior observed in Sr_2IrO_4.

The search of current-controlled materials should focus on AFM Mott insulators with strong structural and/or magnetic distortions in high-Z materials where the role of SOI is significant and electron orbitals are extended: SOI lock magnetic moments to the lattice and the extended orbitals facilitate a strong coupling of current and electron orbitals.

Current-controlled phenomena and materials pose tantalizing prospects for unique functional materials and devices, but a better understanding of them needs to be established first. In particular, the coupling between electrical current and the lattice or orbitals or magnetic moments in this class of materials has yet to be adequately described.

Further Reading

- Yoshinori Tokura, Masashi Kawasaki, and Naoto Nagaosa. Emergent functions of quantum materials. *Nature Physics* 13, 1056 (2017)
- D.N. Basov, R.D. Averitt, and D. Hsieh. Towards properties on demand in quantum materials. *Nature Materials* 16, 1077 (2017)

References

1. Tokura, Y., Kawasaki, M., Nagaosa, N. Emergent functions of quantum materials, *Nature Physics* 13, 1056 (2017)
2. Basov, D.N., Averitt, R.D., Hsieh, D. Towards properties on demand in quantum materials, *Nature Materials* 16, 1077 (2017)
3. Cao, G., DeLong, L. (Eds.) *Frontiers of 4d- and 5d-transition metal oxides.* (Hackensack, NJ: World Scientific, 2013)
4. Witczak-Krempa, W., Chen, G., Kim, Y.B., Balents, L. Correlated quantum phenomena in the strong spin-orbit regime, *Ann. Rev. Condens. Matter Phys.* 5, 57 (2014)
5. Rau, J.G., Lee, E.K.H., Kee, H.Y. Spin-orbit physics giving rise to novel phases in correlated systems: iridates and related materials, *Ann. Rev. Condens. Matter Phys.* 7, 195 (2016)
6. Cao, G., Schlottmann, P. The challenge of spin-orbit-tuned ground states in iridates: a key issues review, *Reports on Progress in Physics* 81, 042502 (2018)
7. Cao, G., Terzic, J., Zhao, H.D., Zheng, H., De Long, L.E., Riseborough, P.S. Electrical control of structural and physical properties via strong spin-orbit interactions in Sr_2IrO_4, *Phys. Rev. Lett.* 120, 017201 (2018)
8. Zhao, H., Hu, B., Ye, F., Hoffmann, C., Kimchi, I., Cao, G. Nonequilibrium orbital transitions via applied electrical current in calcium ruthenates, *Phys. Rev. B* 100, 241104(R) (2019)
9. Cao, G. Topical Review: Towards electrical-current control of quantum states in spin-orbit-coupled matter, *Journal of Physics: Condensed Matter* 32, 423001 (2020)
10. Bertinshaw, J., Gurung, N., Jorba, P., Liu, H., Schmid, M., Mantadakis, D.T., Daghofer, M., Krautloher, M., Jain, A., Ryu, G.H., Fabelo, O., Hansmann, P., Khaliullin, G., Pfleiderer, C., Keimer, B., Kim, B.J. Unique crystal structure of Ca_2RuO_4 in the current stabilized semi-metallic state, *Phys. Rev. Lett.* 123, 137204 (2019)
11. Zhang, J., McLeod, A.S., Han, Q., Chen, X., Bechtel, H.A., Yao, Z., Corder, S.G., Ciavatti, T., Tao, T.H., Aronson, M., et al. Nano-resolved current-induced insulator-metal transition in the Mott insulator Ca_2RuO_4, *Phys. Rev. X* 9, 011032 (2019)
12. Cirillo, C., Granata, V., Avallone, G., Fittipaldi, R., Attanasio, C., Avella, A., Vecchione, A. Emergence of a metallic metastable phase induced by electrical current in Ca_2RuO_4, *Phys. Rev. B* 100, 235142 (2019)
13. Okazaki, R., Kobayashi, K., Kumai, R., Nakao, H., Murakami, Y., Nakamura, F., Taniguchi, H., Terasaki, I. Current-induced giant lattice deformation in the Mott insulator Ca_2RuO_4, *J. Phys. Soc. Japan* 89, 044710 (2020)
14. Guertin, R.P., Bolivar, J., Cao, G., McCall, S., Crow, J.E. Negative differential resistivity in $Ca_3Ru_2O_7$: unusual transport, magnetic coupling in a near-metallic system, *Solid State Comm.* 107, 263 (1998)

15. Cao, G., Crow, J.E., Guertin, R.P., Henning, P.F., Homes, C.C., Strongin, M., Basov, D.N., Lochner, E. Charge density wave formation accompanying ferromagnetic ordering in quasi-one-dimensional $BaIrO_3$, *Solid State Commun.* 113, 657 (2000)
16. Cao, G., Bolivar, J., McCall, S., Crow, J.E., Guertin, R.P. Weak ferromagnetism, metal-to-nonmetal transition, and negative differential resistivity in single-crystal Sr_2IrO_4, *Phys. Rev. B* 57, R11039(R) (1998)
17. Higman, T.K., Miller, L.M, Favaro, M.E., Emanuel, M.A., Hess, K., Coleman, J.J. Room-temperature switching and negative differential resistance in the heterostructure hot-electron diode, *Appl. Phys. Lett.* 53, 1623 (1988)
18. Jeong, D.S., Thomas, R., Katiyar, R.S., Scott, J.F., Kohlstedt, H., Petraru, A., Hwang, C.S. Emerging memories: resistive switching mechanisms and current status, *Rep. Prog. Phys.* 75, 076502 (2012)
19. Chudnovskii, F.A., Pergament, A.L., Stefanovich, G.B., Somasundaram, P., Honig, J.M. Electronic switching in $CuIr_2S_{4-x}Se_x$, *Phys. Status Solidi A* 162, 601 (1997)
20. Pergament, A.L., Stefanovich, G., Velichko, A., Khanin, S.D. Electronic switching and metal-insulator transitions in compounds of transition metals, In: *Condensed matter at the leading edge*, ed M.P. Das, Chapter 1, Pages 1–67. (Nova Science Publishers, 2006)
21. Chudnovskii, F.A., Odynets, L.L., Pergament, A.L., Stefanovich, G.B. Electroforming and switching in oxides of transition metals: the role of metal-insulator transition in the switching mechanism, *J. Solid State Chem.* 122, 95 (1996)
22. Tsen, A.W., Hovden, R., Wang, D., Kim, Y.D., Okamoto, J., Spoth, K.A., Liu, Y., Lu, W., Sun, Y., Hone, J.C., Kourkoutis, L.F., Kim, P., Pasupathy, A.N. Structure and control of charge density waves in two-dimensional $1T\text{-}TaS_2$, *PNAS* 112, 15054 (2015)
23. Okimura, K., Ezreena, N., Sasakawa, Y., Sakai, J. Electric-field-induced multistep resistance switching in planar $VO_2/c\text{-}Al_2O_3$ structure, *Jap. J. Appl. Phys.* 48, 065003 (2009)
24. Okazaki, R., Nishina, Y., Yasui, Y., Nakamura, F., Suzuki, T., Terasaki, I. Current-induced gap suppression in the Mott insulator Ca_2RuO_4, *J. Phys. Soc. Jpn.* 82, 103702 (2013)
25. Kimura, T., Goto, T., Shintani, H., Ishizaka, K., Arima, T., Tokura, Y. Magnetic control of ferroelectric polarization, *Nature* 426, 55 (2003)
26. Tokura, Y., Seki, S., Nagaosa, N. Multiferroics of spin origin, *Rep. Prog. Phys.* 77, 076501 (2014)
27. Frenkel, J. On pre-breakdown phenomena in insulators and electronic semi-conductors, *Phys. Rev.* 54, 647 (1938)
28. Alexander, C.S., Cao, G., Dobrosavljevic, V., Lochner, E., McCall, S., Crow, J.E., Guertin, P.R. Destruction of the Mott insulating ground state of Ca_2RuO_4 by a structural transition, *Phys. Rev. B* 60, R8422 (1999)
29. Nakatsuji, S., Ikeda, S., Maeno, Y. Ca_2RuO_4: New Mott insulators of layered ruthenate, *J. Phys. Soc. Jpn.* 66, 1868 (1997)
30. Braden, M., André, G., Nakatsuji, S., Maeno, Y. Crystal and magnetic structure of Ca_2RuO_4: magnetoelastic coupling and the metal-insulator transition, *Phys. Rev. B.* 58, 847 (1998)
31. Steffens, P., Friedt, O., Alireza, P., Marshall, W.G., Schmidt, W., Nakamura, F., Nakatsuji, S., Maeno, Y., Lengsdorf, R., Abd-Elmeguid, M.M., Braden, M. High-pressure diffraction studies on Ca_2RuO_4, *Phys. Rev. B.* 72, 094104 (2005)
32. Fang, Z., Terakura, K. Magnetic phase diagram of $Ca_{2-x}Sr_xRuO_4$ governed by structural distortions, *Phys. Rev. B.* 64, 020509(R) (2001)
33. Hotta, T., Dagotto, E., Prediction of orbital ordering in single-layered ruthenates, *Phys. Rev. Lett.* 88, 017201 (2001)

34. Lee, J.S., Lee, Y.S., Noh, T.W., Oh, S.-J., Yu, J., Nakatsuji, S., Fukazawa, H., Maeno, Y. Electron and orbital correlations in $Ca_{2-x}Sr_xRuO_4$ probed by optical spectroscopy, *Phys. Rev. Lett.* **89**, 257402 (2002)

35. Jung, J.H., Fang, Z., He, J.P., Kaneko, Y., Okimoto, Y., Tokura, Y. Change of electronic structure in Ca_2RuO_4 induced by orbital ordering, *Phys. Rev. Lett.* **91**, 056403 (2003)

36. Gorelov, E., Karolak, M., Wehling, T.O., Lechermann, F., Lichtenstein, A.I., Pavarini, E. Nature of the Mott transition in Ca_2RuO_4, *Phys. Rev. Lett.* **104**, 226401 (2010)

37. Liu, G.-Q. Spin-orbit coupling induced Mott transition in $Ca_{2-x}Sr_xRuO_4$ ($0 \leq x \leq 0.2$), *Phys. Rev. B.* **84**, 235136 (2011)

38. Nakamura, F., Goko, T., Ito, M., Fujita, T., Nakatsuji, S., Fukazawa, H., Maeno, Y., Alireza, P., Forsythe, D., Julian, S.R. From Mott insulator to ferromagnetic metal: a pressure study of Ca_2RuO_4, *Phys. Rev. B.* **65**, 220402(R) (2002)

39. Zegkinoglou, I., Strempfer, J., Nelson, C.S., Hill, J.P., Chakhalian, J., Bernhard, C., Lang, J.C., Srajer, G., Fukazawa, H., Nakatsuji, S., Maeno, Y., Keimer, B. Orbital ordering transition in Ca_2RuO_4 observed with resonant X-ray diffraction, *Phys. Rev. Lett.* **95**, 136401 (2005)

40. Nakatsuji, S., Maeno, Y. Quasi-two-dimensional Mott transition system $Ca_{2-x}Sr_xRuO_4$, *Phys. Rev. Lett.* **84**, 2666 (2000)

41. Snow, C.S., Cooper, S.L., Cao, G., Crow, J.E., Fukazawa, H., Nakatsuji, S., Maeno, Y. Pressure-tuned collapse of the Mott-like state in $Ca_{n+1}Ru_nO_{3n+1}$ ($n = 1,2$): Raman Spectroscopic Studies, *Phys. Rev. Lett.* **89**, 226401 (2002)

42. Cao, G., McCall, S., Shepard, M., Crow, J.E., Guertin, R.P. Magnetic and transport properties of single-crystal Ca_2RuO_4: relationship to superconducting Sr_2RuO_4, *Phys. Rev. B* **56**, R2916(R) (1997)

43. Qi, T.F., Korneta, O.B., Parkin, S., DeLong, L.E., Schlottmann, P., Cao, G. Negative volume thermal expansion via orbital and magnetic orders in $Ca_2Ru_{1-x}Cr_xO_4$, *Phys. Rev. Lett.* **105**, 177203 (2010)

44. Qi, T.F., Korneta, O.B., Parkin, S., Hu, J., Cao, G. Magnetic and orbital orders coupled to negative thermal expansion in Mott insulators $Ca_2Ru_{1-x}M_xO_4$ (M = Mn and Fe), *Phys. Rev. B.* **85**, 165143 (2012)

45. Fujiyama, S., Ohsumi, H., Komesu, T., Matsuno, J., Kim, B.J., Takata, M., Arima, T., Takagi, H. Two-dimensional Heisenberg behavior of $J_{eff} = 1/2$ isospins in the paramagnetic state of the spin-orbital Mott insulator Sr_2IrO_4, *Phys. Rev. Lett.* **108**, 247212 (2012)

46. Dai, J.X., Calleja, E., Cao, G., McElroy, K. Local density of states study of a spin-orbit-coupling induced Mott insulator Sr_2IrO_4, *Phys. Rev. B* **90**, 041102(R) (2014)

47. Ye, F., Chi, S.X., Chakoumakos, B.C., Fernandez-Baca, J.A., Qi, T.F., Cao, G. Magnetic and crystal structures of Sr_2IrO_4: a neutron diffraction study, *Phys. Rev. B* **87**, 140406(R) (2013)

48. Boseggia, S., Walker, H.C., Vale, J., Springell, R., Feng, Z., Perry, R.S., Moretti Sala, M., Rønnow, H.M., Collins, S.P., McMorrow, D.F. Locking of iridium magnetic moments to the correlated rotation of oxygen octahedra in Sr_2IrO_4 revealed by X-ray resonant scattering, *J. Phys.: Condens. Matter* **25**, 422202 (2013)

49. Torchinsky, D.H., Chu, H., Zhao, L., Perkins, N.B., Sizyuk, Y., Qi, T., Cao, G., Hsieh, D. Structural distortion-induced magnetoelastic locking in Sr_2IrO_4 revealed through nonlinear optical harmonic generation, *Phys. Rev. Lett.* **114**, 096404 (2015)

50. Jackeli, G., Khaliullin, G. Mott insulators in the strong spin-orbit coupling limit: from Heisenberg to a quantum compass and Kitaev models, *Phys. Rev. Lett.* **102**, 017205 (2009)

51. Wang, J.C., Aswartham, S., Ye, F., Terzic, J., Zheng, H., Haskel, D., Chikara, S., Choi, Y., Schlottmann, P., Custelcean, R., Yuan, S.J., Cao, G. Decoupling of the antiferromagnetic and insulating states in Tb-doped Sr_2IrO_4, *Phys. Rev. B* **92**, 214411 (2015)

52. Ye, F., Hoffmann, C., Tian, W., Zhao, H., Cao, G. Pseudospin-lattice coupling and electric control of the square-lattice iridate Sr_2IrO_4, *Phys. Rev. B* 102, 115120 (2020)
53. Ge, M., Qi, T.F., Korneta, O.B., De Long, L.E., Schlottmann, P., Crummett, W.P., Cao, G. Lattice-driven magnetoresistivity and metal-insulator transition in single-layered iridates, *Phys. Rev. B* 84, 100402(R) (2011)
54. Cao, G., Xin, Y., Alexander, C.S., Crow, J.E., Schlottmann, P., Crawford, M.K., Harlow, R.L., Marshall, W. Anomalous magnetic and transport behavior in the magnetic insulator $Sr_3Ir_2O_7$, *Phys. Rev. B* 66, 214412 (2002)
55. Lovesey, S.W., Khalyavin, D.D., Manuel, P., Chapon, L.C., Cao, G., Qi, T.F. Magnetic symmetries in neutron and resonant x-ray Bragg diffraction patterns of four iridium oxides, *J. Phys. Condens. Matter* 24, 496003 (2012)
56. Kim, J.W., Choi, Y., Kim, J., Mitchell, J.F., Jackeli, G., Daghofer, M., van den Brink, J., Khaliullin, G., Kim, B.J. Dimensionality driven spin-flop transition in layered iridates, *Phys. Rev. Lett.* 109, 037204 (2012)

Part 3

Single-Crystal Synthesis

Chapter 6
Single-Crystal Synthesis

6.1 Overview

High-quality, bulk single crystals are essential for definitive studies of fundamental properties as well as successful integration of their properties into state-of-the-art device structures. Investigators who first synthesize high-quality crystals of novel materials are in an optimal position to discover their frequently surprising properties and, ultimately, to lead efforts to apply them in advanced technologies. However, growing single crystals of complex materials is often a great challenge. This is certainly true for $4d$- and $5d$-transition metal oxides, as they tend to form incongruently, and have high vapor pressure and high melting points.

Two crystal growth techniques are commonly used for transition metal oxides—flux and floating-zone techniques. Both techniques have advantages and disadvantages (**Table 6.1**), but neither is adequate for synthesis of a wide range of materials. Combining the two techniques makes it possible to grow single crystals of almost all stable materials. A large number of single crystals of $4d$- and $5d$-transition metal oxides grown by either the flux or floating-zone technique are listed in Table 1.4.

Crystal structures of most $4d$- and $5d$-transition metal oxides are inherently distorted. Because of the strong spin-orbit interactions (SOI), local structural distortions often dictate the ground states of these materials. Indeed, as discussed in Chapter 1, the absence of many theoretically predicted states could be, in part, attributed to local distortions. In response to this challenge, an innovative, "field-altering" technique was recently proposed [1], in which an applied magnetic field aligns magnetic moments and, through strong SOI and magnetoelastic coupling, alters or "corrects" crystal structures at high temperatures. A proof-of-concept work demonstrates that a field-altering technology is highly effective in the case of materials with a delicate interplay between SOI and comparable Coulomb correlations. Moreover, new quantum states arise whenever such competing interactions conspire to provoke unusually large susceptibilities to small stimuli.

Similar to cooking, the art of crystal growth is a skill acquired through practice. Each material requires its own unique growth conditions that must be optimized by trial and error. This chapter is intended to provide some basic aspects of both flux and floating-zone techniques, and offer a few general remarks on crystal growth of $4d$- and $5d$-transition metal oxides. The field-altering technology for synthesis of certain

Physics of Spin-Orbit-Coupled Oxides. Gang Cao and Lance E. DeLong, Oxford University Press (2021). © Gang Cao and Lance E. DeLong.
DOI: 10.1093/oso/9780199602025.003.0006

Table 6.1 *Techniques for Single-Crystal Growth*

Techniques	Pros	Cons
Floating-Zone	Large crystals No crucible contamination Atmosphere control Pressure control (up to 150 bar) Fast growth rate (0.1–50 mm/hr) High temperature (> 2000 °C)	Inhomogeneity/intergrowth Incongruent melting Ineffective for a number of compounds, e.g., iridates Sensitive to control parameters Tedious preparation Expensive
Flux	Homogeneity Natural crystal shape or habit Suitable for most materials	Relatively small crystals Possible contamination

spin-orbit-coupled oxides is also described as a promising new tool. Fundamental principles and general practical issues encountered during crystal growth have been thoroughly discussed in a great body of literature over many decades [2–4 and references therein].

6.2 Flux Technique

Flux growth is the most common, versatile technique used for crystal growth. A flux consists of one or multiple solvents. A solution is a combination of a flux and a solute material to be grown. The most critical requirement to initiate crystal growth is to establish supersaturation or a nonequilibrium state in the solution in which the solute concentration is higher than that in the equilibrium state. As schematically illustrated in **Fig. 6.1**, there are three process steps required to achieve supersaturation: solvent evaporation (arrow 1), temperature gradient (arrow 2), and slow cooling (arrow 3). Often, all three processes are simultaneously employed to attain crystal growth.

The flux technique is particularly suitable for materials that have some of the following unfavorable features for crystal growth: (1) incongruent melting (peritectic decomposition below the melting point), (2) high melting point, (3) high vapor pressure, and (4) formation of nonstoichiometric phases due to a highly volatile element. Most $4d$- and $5d$-transition metal oxides exhibit at least some of these difficulties. A key drawback of this technique is possible inclusion of flux impurities in the crystal. Alternatively, a self-flux, which contains an element needed in the crystal to be grown, is often used to mitigate the problem. Many single crystals of $4d$- and $5d$-transition metal oxides are grown from a self-flux; for example, a combination of $SrCl_2$ and $SrCO_3$ is used as a self-flux for growth of materials containing Sr, such as $SrRuO_3$, $Sr_5Rh_4O_{12}$, Sr_2IrO_4, and so on. If there is no obvious self-flux for certain compounds, such as $RERhO_3$ (RE = rare earth ion), a K_2CO_3 flux could be used for crystal growth of these materials [5].

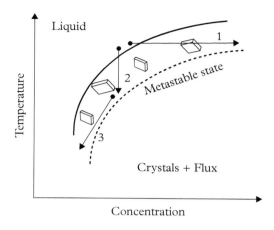

Fig. 6.1 *Schematic illustration of three processes to achieve supersaturation: 1. solvent evaporation, 2. temperature gradient, and 3. slow cooling.*

Ideally, a flux should have high solubility, low viscosity, low volatility, low vapor pressure, and low corrosion. In addition, it should be easily dissolved in water so that the flux can be separated from the crystal product. In practice, it is difficult to find a flux that has all these favorable characteristics. One should be mindful that high solubility requires similar bonding (e.g., ionic bonding, covalent bonding): many transition metal oxides tend to be ionically bonded; therefore single crystals of these materials are often grown from a flux that is made of ionic compounds. It is also important to select a flux that avoids formation of unwanted phases via unwanted reactions between solute and flux. For example, PbO and/or PbF_2, which are common solvents for oxides, should be avoided as a flux for crystal growth of $BaRuO_3$ or $BaIrO_3$ because an undesired $BaPbO_3$ phase can readily form.

It has long been recognized that a fluoride alone or mixed with an oxide (e.g., PbO-PbF_2) helps reduce the viscosity of the flux, which can generate larger crystals [2]. Adding B_2O_3 to a flux could reduce nucleation, thus leading to fewer but larger crystals. This occurs because an enlarged metastable region reduces further nucleation once crystals start to form [6].

Stability during crystal growth is another important aspect, since smooth surfaces of crystals are a result of stable growth. Dendrites or depressions on crystal surfaces are indicative of unstable growth because supersaturation tends to be higher at the corners and edges of crystals than at the centers of facets. Flux inclusions in the crystals are also a sign of unstable growth.

Single crystals of many $4d$- and $5d$-transition metal oxides are often grown at high temperatures (> 1200°C); therefore platinum crucibles are relatively desirable because they are more chemically inert. However, platinum crucibles are expensive (Pt is one of the nine least-abundant elements on the Earth's crust). Other less-expensive materials such as alumina or zirconia are also commonly used to fabricate crucibles for crystal

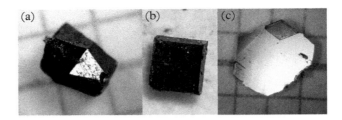

Fig. 6.2 *Representative flux-grown single crystals of (a) hexagonal BaIrO$_3$, (b) orthorhombic Ca$_3$Ru$_2$O$_7$, and (c) honeycomb (Na$_{1-x}$Li$_x$)$_2$IrO$_3$.*

growth of these materials. These crucibles can withstand temperatures up to 1700°C, but the formation of inclusions of undesired elements such as Al in the crystals could be an issue in some cases. Corrosion by reactions with the crucible should be avoided as well. For example, alumina crucibles should not be used for crystal growth that includes PbO. Creeping is another issue for crystal growth if the crucible is not properly selected. For this reason, alumina crucibles should not be used for crystal growth of materials bearing La, for example.

Furnaces used for crystal growth must have high temperature stability. Uncontrolled temperature fluctuations often result in either very small crystals or no crystals at all. As such, a furnace with a larger thermal mass is always more desirable.

The flux technique allows a more natural crystal growth than the floating-zone technique discussed later. In particular, the characteristic external shape of a crystal (i.e., the "crystal habit") is necessarily determined by the crystal structure (e.g., **Fig. 6.2**). Note that the longest axis of a unit cell always corresponds to a shortest dimension of a flux-grown crystal because it is harder for the crystal to expand along the direction. This explains the common, plate-like shape of quasi-two-dimensional crystals, such as Ca$_2$RuO$_4$ and Sr$_2$IrO$_4$, in which the thinnest dimension of the crystals corresponds to the longest *c* axis.

Single crystals of most 4*d*- and 5*d*-transition oxides studied thus far can be grown using the flux technique. Some of them (e.g., iridates Sr$_2$IrO$_4$, Na$_2$IrO$_3$) are large enough for neutron studies despite a large neutron absorption cross-section for isotopes of Ir [7,8].

6.3 Optical Floating-Zone Technique

The optical floating-zone technique has been a successful method for crystal growth for decades, and is effective for growing crystals of many 3*d*- and some 4*d*-transition metal oxides [3]. It complements the flux technique (see **Table 6.1**) in part because this technique can generate larger and purer crystals of certain materials, which is an important consideration for neutron studies and experiments that require ultra-high-purity samples, as is the case for Sr$_2$RuO$_4$, where trace impurities readily suppress superconductivity (see **Fig. 6.3**).

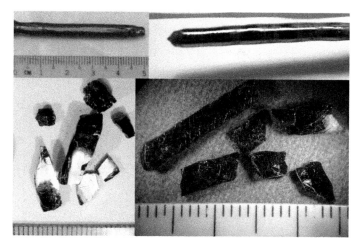

Fig. 6.3 *Representative floating-zone-grown crystals of Sr ruthenates. Note that the two long crystal rods in the upper panels are what the crystals look like right after coming out of the floating-zone furnace. They often consist of many smaller single crystals such as those shown in the lower panels.*

The floating-zone technique was first developed at Bell Laboratories and commercialized by Siemens in the 1950s. This new technology, which ruled out possible crucible contamination, permitted synthesis of ultrapure silicon single crystals, which was a critical requirement for the development of many semiconductor devices. More recently, the floating-zone technique was improved by using halogen or xenon lamps and ellipsoidal mirrors for crystal growth of a wide range of metals and oxides (**Fig. 6.4**). Float-zoned crystals have proved to be of great fundamental and technological significance, and floating-zone furnaces are now used at a large number of research universities and national laboratories throughout the world. In recent years, halogen or xenon lamps have been replaced by much more compact laser diodes, which eliminated the need for large ellipsoidal mirrors. Optical floating-zone furnaces available today can operate at 2000°C or higher with appropriate atmospheric controls.

In an optical floating-zone furnace, the light from the lamps or the laser diodes is tightly focused on a vertical feed rod (that supports the growing crystal) to create a molten zone that slowly moves vertically along the rod during crystal growth. The surface tension of the molten zone provides support against gravity so that the molten zone "floats" without a crucible (**Fig. 6.4**).

The crystal growth involves two important steps: feed rod preparation and crystal growth. The overall growth process—especially feed rod preparation—can be tedious and time-consuming.

A feed rod is made of well-ground, fine powders of suitable chemical feed stock that is usually pressurized and then fired at high temperatures. The quality of the rod is often measured by its straightness, density, and homogeneity. A slight bend in the rod causes a rotational wobble, which significantly degrades the stability of the molten zone (**Fig. 6.4a**).

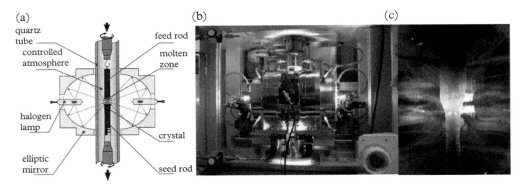

Fig. 6.4 *Floating-zone technique: (a) a schematic illustration of the floating zone. (b) A floating-zone furnace in operation. (c) A snapshot of the molten zone during crystal growth.*

A porous or inhomogeneous feed rod may produce bubbles in the molten zone, or cause it to penetrate into the feed rod, which will also strongly degrade the stability of the molten zone. High-temperature sintering of the rod can help reduce porousness, but this does not always work, especially if the rod is not made of well-ground fine powder. If a feed rod contains a volatile element, it is often made nonstoichiometric with an increased amount of the volatile element. For example, RuO_2 is highly volatile above $1200°C$; therefore, feed rods are often made Ru-rich to compensate the Ru loss during growth. This practice is often used for crystal growth of ruthenates, such as $Ca_3Ru_2O_7$, Sr_2RuO_4, etc. In short, a high-quality feed rod is a critical requirement for successful floating-zone crystal growth.

Crystal growth usually requires control of a number of parameters: growth rate, rod rotation rate, temperature, and atmosphere. The growth rate depends on the growth kinetics of the material (e.g., congruent or incongruent melting). The growth rate normally ranges from 1 to 50 mm/hour [3]. The growth rate for some ruthenates lies between 6 and 35 mm/hour. An optimal growth rate should help generate large crystals and avoid cracks and undesired phases. Generally, a slower growth rate generates larger crystals.

A homogeneous molten zone is, to a certain extent, established and maintained by the rotation of the feed rod (see **Fig. 6.4a**). In general, a higher rotation rate is more desirable because it increases convection, which can help generate a more stable molten zone.

Maintenance of an optimal temperature gradient in the rod is important. It is affected by a number of factors such as the chamber atmosphere, the focus of the lamps, and material properties (e.g., thermal expansion). An excessive temperature gradient may cause cracks due to thermal stress; on the other hand, a sharp temperature gradient helps reduce the molten zone, and a smaller zone is a more stable zone.

Most optical floating-zone furnaces offer atmospheric control (e.g., vacuum, gas flow, application of modest pressure). Many transition metal oxides are grown in air, oxygen, or a mixture of oxygen and inert gases. Flowing argon favors growth of metallic materials. Application of pressure helps reduce vaporization of volatile elements from the rod and affects (or controls) the chemical composition of the crystals. However, the stability of the molten zone could decrease with increasing pressure due to possible bubble formation. In addition, the melting point of the rod increases with increasing pressure, thus requiring more power for crystal growth. Cracks in the crystal may occur when applied pressure is too high. Less favorable growth conditions generally cause inclusions of unwanted phases (e.g., $Sr_4Ru_3O_{10}$ mixing with $SrRuO_3$), inhomogeneities in the crystals, and other problems.

The floating-zone technique requires a relatively large amount of chemicals for each growth. The mass of a minimum size of a feed rod is at least several grams; therefore, the costs for chemicals could be high and, in some cases, unaffordable or unstainable if the crystals contain very expensive elements. For this reason, the floating-zone technique is not commonly used for crystal growth of iridates.

6.4 Field-Altering Technology

Recently, it has been found that new quantum states in single crystals of 4*d*- and 5*d*-electron-based oxides can be obtained through structurally "altering" materials via application of magnetic field during crystal growth. The key mechanism is that magnetic field aligns magnetic moments and, through strong SOI and magnetoelastic coupling, alters or "corrects" crystal structures at high temperatures, as schematically illustrated in **Fig. 6.5a**. Such a "field-altering" technology works particularly well for materials that are governed by strongly competing fundamental interactions [1], which lead to multiple, nearly degenerate states.

Field-altering technology directly addresses a major challenge to current research on compounds with 4*d*- and 5*d*-transition elements: a great deal of theoretical work that predicts exotic states for correlated and spin-orbit-coupled oxides has thus far met very limited experimental confirmation. It has become increasingly clear, as discussed in previous chapters, that the discrepancies are due chiefly to the extreme susceptibility of these materials to structural distortions and disorder. The application of magnetic field during materials growth at high temperatures (**Fig. 6.5a**) can alter distortions/disorder and thereby provoke unusually strong responses in physical properties, even stabilizing a new ground state (see, for example, **Figs. 6.5b** and **6.5c**). A magnetic field applied during crystal growth exerts a torque on magnetic moments present in the crystal, which via strong SOI can change the bond angles and the overlap matrix elements of the orbitals and, consequently, the physical properties.

Strong magnetic fields (e.g., on the order of several Tesla) have yet to be applied for crystal synthesis of 4*d*- and 5*d*-materials. Nevertheless, important proof-of-concept results demonstrate that drastic changes in physical properties take place after application of a very weak magnetic field no stronger than 0.06 Tesla. This appears utterly inconsistent with conventional thermodynamics, given that even an extremely strong magnetic field (e.g., 45 Tesla ~ 4 meV) seems inconsequential in comparison to the energetics of chemical reactions, as magnetic contributions to the Gibbs free enthalpy are too small to be significant [9]. Indeed, previous literature on applying magnetic fields during crystal

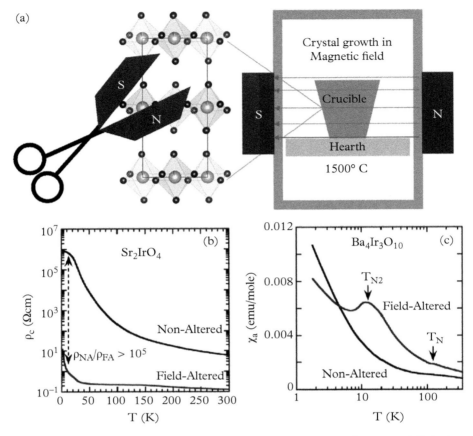

Fig. 6.5 *Field-altering technology: (a) a schematic for field-altering a crystal structure (left) during crystal growth in the molten zone sandwiched between the two magnets (right). Contrasting physical properties of two exemplary materials to highlight field-altering effects: (b) the c-axis resistivity ρ_c of Sr_2IrO_4 is five orders of magnitude smaller in the field-altered (FA) crystal than in the non-altered (NA) counterpart. (c) A comparison of the a-axis magnetic susceptibility χ_a of $Ba_4Ir_3O_{10}$ between the non-altered crystal that features a quantum liquid and the field-altered crystal that becomes an antiferromagnet with two magnetic transitions [1].*

growth of silicon [10–12] and protein crystals [13] were mostly limited to only changing surface striations of the grown crystals [10].

In the proof-of-concept setup, two permanent magnets, each of which is 8" in diameter and generates 1.4 Tesla, are aligned and mounted on the two opposite sides of a 1500°C box furnace (**Fig. 6.5a**). Because the axial magnetic field of a permanent bar magnet decays with distance d as $1/d^3$, the actual strength of the magnetic field inside the furnace chamber is in the range of 0.02–0.06 Tesla, sensitively depending on the exact location. The magnets used are NdFeB magnets. The Curie temperature of this class of magnets is between 320 and 400°C. The magnetic field is strongest below 80°C; above 80°C, the magnet loses its field strength by about 0.11% for each degree (°C) increase. Therefore, the temperature of the magnets needs to be closely monitored, especially when the furnace temperature is high.

In the following section, the structural and physical properties of field-altered and non-altered samples are compared. The field-altered crystals were grown using the proof-of-concept setup illustrated in **Fig. 6.5a**.

6.4.1 $Ba_4Ir_3O_{10}$: From Quantum Liquid to Correlated Antiferromagnet

As discussed in Chapter 3, the magnetic insulator $Ba_4Ir_3O_{10}$ is a novel quantum liquid. The novelty of the state is that magnetic frustration occurs in an unfrustrated square lattice of Ir_3O_{12} trimers composed of face-sharing IrO_6 octahedra (**Figs. 6.6d–6.6e**). It is these trimers that form the basic magnetic unit and play a crucial role in frustration. Specifically, the direct (Ir-Ir) and superexchange (Ir-O-Ir) interactions in the trimers interfere such that the middle Ir ion in a trimer is only very weakly linked to the two neighboring Ir ions. Such "weak links" generate an effective one-dimensional system with zigzag chains or Luttinger liquids along the *c* axis.

The intricate balance of interactions is fundamentally changed in field-altered $Ba_4Ir_3O_{10}$ crystals. Compared to those of the non-altered sample, field-altered single crystals exhibit a significant elongation of the *b* axis with only slight changes in the *a* and *c* axes. As a result, the unit cell volume V increases by up to 0.54% at 350 K (see **Figs. 6.6a–6.6c**). Both the Ir-Ir bond distance within each trimer and the Ir-O-Ir bond angle between trimers increase substantially, as illustrated in **Figs. 6.6f** and **6.6g**.

Field-altering not only destroys the peculiar quantum liquid presented by the non-altered samples, but also stabilizes a robust, long-range magnetic order. As shown in **Fig. 6.7a**, two magnetic anomalies occur at Néel temperatures $T_N = 125$ K and $T_{N2} = 12$ K in the field-altered sample, sharply contrasting with the magnetic behavior of the non-altered sample. Consequently, the absolute values of the Curie-Weiss temperatures θ_{CW} are considerably reduced and become comparable to $T_N = 125$ K for the field-altered sample (**Figs. 6.7b** and **6.7c**). The corresponding frustration parameter f = $|\theta_{CW}|/T_N$ is accordingly reduced to a value less than 3, which is a drastic drop from the average value of 2000 in the non-altered sample, and indicates a complete removal of frustration. The long-range magnetic order is corroborated by a metamagnetic transition observed in the isothermal magnetization M(H) along the *a* axis at a critical field $\mu_oH_c = 2.5$ T (**Fig. 6.7d**). Metamagnetism often occurs in canted antiferromagnetic (AFM) states in quasi-two-dimensional lattices.

The heat capacity, which measures bulk effects, confirms the AFM order. In particular, the low-temperature linearity of the heat capacity C(T) (**Fig. 6.7e**), which characterizes the gapless excitations in the non-altered sample, is replaced by the T^3-dependence in the field-altered sample, which is anticipated for an insulating antiferromagnet. Along with the linearity of C(T), the sharp upturn in C(T) at $T^* = 0.2$ K in the non-altered

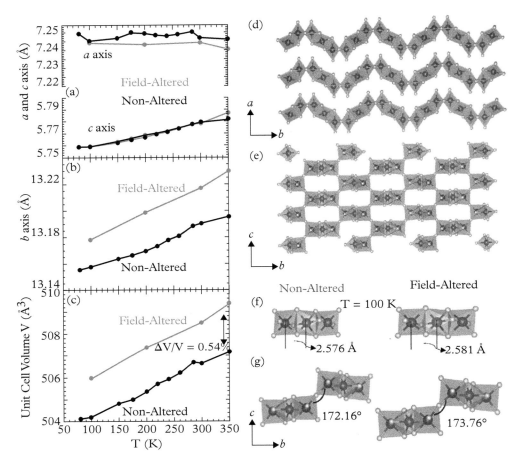

Fig. 6.6 $Ba_4Ir_3O_{10}$: *Structural properties of the field-altered and non-altered single crystals: (a) the temperature dependence of the lattice parameters for the a and c axes; (b) the b axis, and (c) the unit cell volume V. (d) and (e) show the crystal structure in the ab- and bc-planes, respectively. (f) The Ir-Ir bond distance within a trimer. (g) The Ir-O-Ir bond angle between corner-sharing trimers (the marked values are for 100 K) [1].*

sample also disappears in the field-altered sample. Further, as temperature rises, two anomalies occur at $T_{N2} = 12$ K (**Fig. 6.7f**) and $T_N = 125$ K (**Fig. 6.7g**), respectively, supporting the existence of long-range magnetic order indicated by the magnetic data in **Figs. 6.7a–6.7d**. These changes clearly illustrate that the ground state of $Ba_4Ir_3O_{10}$ is fundamentally changed by field altering!

In summary, the quantum liquid state present in non-altered $Ba_4Ir_3O_{10}$, which is attributed to the reduced intra-trimer exchange and weakly coupled one-dimensional chains along the c axis, is supplanted in field-altered $Ba_4Ir_3O_{10}$ by a strongly AFM state stabilized by three-dimensional correlations.

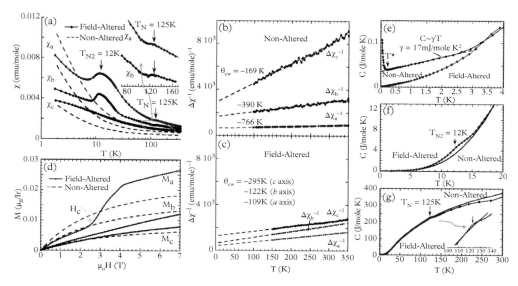

Fig. 6.7 $Ba_4Ir_3O_{10}$: *Physical properties of the field-altered and non-altered single crystals: the temperature dependence of the a-, b-, and c-axis values of (a) the magnetic susceptibility $\chi(T)$ for the field-altered (solid dots) (inset: zoomed-in χ near T_N), and the non-altered (dashed lines) samples, and (b) $\Delta\chi^{-1}$ for the non-altered single crystal, and (c) for the field-altered single crystal. (d) The isothermal magnetization M(H) at 1.8 K for the field-altered (solid lines) and the non-altered (dashed lines) samples. (e) The temperature dependence of the heat capacity C(T) for the field-altered and the non-altered samples at the lowest temperatures, (f) intermediate temperatures, and (g) high temperatures. Inset in (g): the zoomed-in C(T) near T_N [1].*

6.4.2 Ca_2RuO_4: From Collinear Antiferromagnet to Weak Ferromagnet

As discussed in Chapter 4, the AFM insulator Ca_2RuO_4 undergoes a metal-insulator transition at $T_{MI} = 357$ K, which marks a concomitant, violent structural transition with a severe rotation and tilt of RuO_6 octahedra and an AFM transition at a considerably lower Néel temperature $T_N = 110$ K. Extensive investigations of this system have established that quantum effects are intimately coupled to lattice perturbations.

The crystal structure of Ca_2RuO_4 is significantly field altered, becoming less distorted. In particular, the structural transition is reduced by about 25 K in the field-altered sample [1]. Indeed, the *a*-axis electrical resistivity ρ_a of the field-altered sample shows a much lower metal-insulator transition T_{MI} at 324 K, 31 K lower than T_{MI} for the non-altered sample, as seen in **Fig. 6.8a**. The magnetic properties of the field-altered sample behave very differently from the non-altered sample. In particular, the *a*-axis magnetic susceptibility χ_a of the field-altered sample shows a ferromagnetic-like behavior with a magnetic transition at T_N = 135 K, in sharp contrast to the behavior of the non-altered sample (see **Fig. 6.8b**). Moreover, a large hysteresis in the data for χ_a is observed in the field-altered sample (**Fig. 6.8b**, inset), as expected for a ferromagnet or weak ferromagnet. This behavior is also consistent with a metamagnetic transition observed in the isothermal magnetization M(H) at $\mu_o H_c$ = 2.4 T, as illustrated in **Fig. 6.8c**. The changes registered in magnetic data are also in accordance with changes in the low-temperature heat capacity C(T). For an insulating antiferromagnet, C(T) ~ $(\alpha + \beta) T^3$, in which the first term α and the second term β are associated with magnon and phonon contributions to C(T), respectively. Here, C(T) shows a significant slope change defined by $(\alpha + \beta)$ in the plot of C/T vs. T^2 in **Fig. 6.8d**. This slope change clearly points to the emergent magnetic state being distinctly different from the native AFM state.

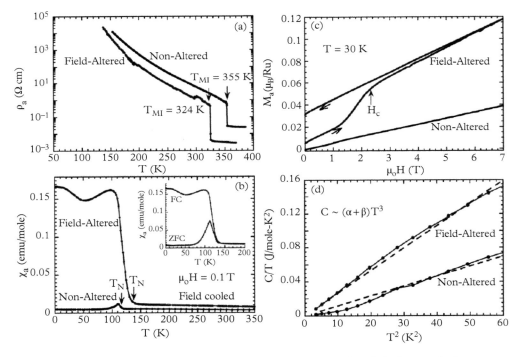

Fig. 6.8 Ca_2RuO_4: *Physical properties of the field-altered and non-altered single crystals: the temperature dependence of (a) the a-axis electrical resistivity ρ_a and (b) the a-axis magnetic susceptibility χ_a(T) at $\mu_o H = 0.1$ T. Inset: χ_a(T) for the field-cooled (FC) and zero-field-cooled (ZFC) sequences for the field-altered sample. (c) The a-axis isothermal magnetization M_a(H) at 30 K. (d) The low-temperature heat capacity C(T) plotted as C/T vs. T^2 [1].*

6.4.3 Sr$_2$IrO$_4$: Toward Novel Superconductivity

As discussed in Chapter 2, Sr$_2$IrO$_4$ is an archetype of the spin-orbit-coupled Mott insulator. The field-altered structure is more expanded and less distorted (i.e., the Ir-O-Ir bond angle becomes larger; see **Figs. 6.9a** and **6.9b**), and the AFM transition T$_N$ is suppressed by an astonishing 90 K (**Fig. 6.9c**). The isothermal magnetization is reduced by 50% and is much less "saturated" than non-altered Sr$_2$IrO$_4$ [1].

Magnetic changes are clearly reflected in Raman scattering data. One-magnon Raman scattering measures the anisotropy field that pins the magnetic moment orientation. It broadens with increasing temperature and vanishes at T$_N$. At 10 K, this peak in the non-altered Sr$_2$IrO$_4$ occurs near 18 cm^{-1} (**Fig. 6.9d**) but is absent in field-altered Sr$_2$IrO$_4$ over the measured energy range. The conspicuous disappearance of the peak clearly indicates that the anisotropy field is drastically reduced and, consequently, the one-magnon peak is either completely removed or suppressed to energies below the cutoff of 5.3 cm^{-1} (0.67 meV) in the field-altered sample. On the other hand, two-magnon scattering remains essentially unchanged [1].

Furthermore, the resistivity is reduced by up to five orders of magnitude and shows a nearly metallic behavior at high temperatures in field-altered Sr$_2$IrO$_4$ (**Figs. 6.9e** and **6.9f**)! Also note that there is an anomaly corresponding to T$_N$ = 150 K (**Fig. 6.9e**, inset), indicating a close correlation between the transport and magnetic properties that is noticeably absent in non-altered Sr$_2$IrO$_4$ (**Figs. 6.9c, 6.9e**, and **6.9f**) [1]. This is consistent with the fact that the drastically improved conductivity (**Figs. 6.9e** and **6.9f**) is accompanied by a correspondingly weakened magnetic state (**Figs. 6.9c** and **6.9d**).

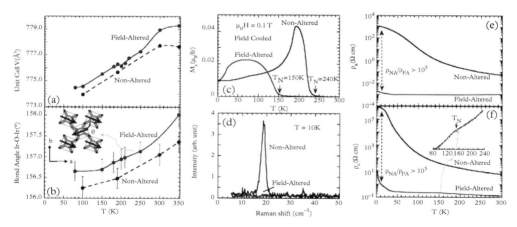

Fig. 6.9 *Sr$_2$IrO$_4$: Structural and physical properties of the field-altered and non-altered single crystals: the temperature dependence of (a) the unit cell volume V, (b) the Ir-O-Ir bond angle θ (inset illustrates θ), and (c) the a-axis magnetization M$_a$ at μ$_0$H = 0.1 T. (d) The Raman single-magnon peak at 10 K. The temperature dependence of (e) the a-axis resistivity ρ$_a$ and (f) the c-axis resistivity ρ$_c$ for Sr$_2$IrO$_4$. Inset: The anomaly in ρ$_a$ at T$_N$ for the field-altered crystal [1].*

6.4.4 Proposed Theoretical Mechanism for Field Altering

The tiny, 200-gauss fields that are sufficient for field-altering indicate a magnetic Zeeman energy scale of only 0.002 meV for a spin-1/2, which is orders of magnitude smaller than typical electronic energy scales. It is not immediately clear how any familiar mechanism could possibly instigate field altering with such small energy scales. Nevertheless, the following presents a possible proof-of-concept mechanism that may admit such an effect.

First, consider the effects of magnetic fields on crystal growth observed in other systems. Crystallization of proteins has been shown to benefit from large applied magnetic fields on the order of a few Tesla or more [13]. The mechanisms for this effect are thought to rely on increased viscosity and more uniform magnetic alignment within the high-temperature fluid. However, the energy landscape for protein crystallization is dominated by ultraweak van der Waals interactions, such that magnetic fields of only a few Tesla can have a direct impact on the preferred alignments and associated crystallization energies. In contrast, the transition metal oxides of interest here are crystallized via covalent and ionic bonding energies that are many orders of magnitude higher. A first distinction between the present phenomena of interest and previous work is the ratio of effective magnetic field strengths to crystallization energies.

A second distinction is in the direct effect of growth in a magnetic field. In the literature on semiconductor crystal growth as well as in the protein crystallization cases, it has been shown that magnetic fields enable higher quality crystals or larger crystals [9–13]. These effects arise from the interface between the high temperature melt and the growing crystal, thus are readily understood in terms of the magnetic field effects in the molten fluid properties. In contrast, the applied field preserves the formation of large bulk crystals but modifies the resulting crystal structure, including its electronic properties.

The foregoing observations suggest that, while the high temperature magnetohydrodynamic effects observed elsewhere are expected to also play a role, an additional, qualitatively new mechanism is necessary to explain field altering. The question to be addressed here is this: What energetic mechanism could allow a weak magnetic field to change the local bond angles by even a few percent, thus inducing the associated dramatic changes in the electronic and magnetic properties discussed earlier? The following qualitatively new mechanism for field altering is based on the combination of (1) strong spin-orbit or magnetoelastic couplings and (2) magnetic frustration.

Consider the low-energy theory of magnetic insulators, which can be expressed in terms of effective spin-orbit-coupled spin-1/2 degrees of freedom represented by the Pauli matrices τ^{α}. The interaction among these degrees of freedom can be written schematically as

$$H = \mathcal{J}_{r,r'}^{\alpha\alpha'} \sum_{r,r'} \tau_r^{\alpha} \tau_{r'}^{\alpha'}$$

The coupling of magnetic field to the crystal structure appears in two ways: in the spin-orbital wavefunction content of τ^{α}, and in the magnetoelastic dependence of the interaction $\mathcal{J}$ on the spatial positions r and r' [1].

Such Hamiltonians can exhibit strong magnetic frustration and associated near-degeneracies among exponentially many quantum states. This can occur even when the underlying magnetic lattice has no geometric frustration; rather the frustration can arise purely from spin-orbit interactions. As a proof-of-concept example, consider the square-lattice, quantum compass model [14]

$$H_c = \mathcal{J}_r^h \sum_{h-\text{bond}} \tau_r^x \tau_{r+x_0}^x + \mathcal{J}_r^v \sum_{v-\text{bond}} \tau_r^y \tau_{r+y_0}^y$$

where h-bonds and v-bonds refer to horizontal and vertical bonds, respectively, with the vectors along the bond directions being x_0 and y_0, respectively. This interaction term has been argued to arise in Sr_2IrO_4, and is also expected to arise via symmetry considerations in Ca_2RuO_4.

The Hamiltonian H_c possesses fine-tuned symmetries that give it unusual properties, which here also include unusual changes in the crystal and electronic structures [14]. First, every eigenstate of this Hamiltonian, including the ground state, is degenerate with an enormous number of other states. This yields a sequence of degenerate manifolds, each of which is exponentially large in system size (2^L). Second, 1D domain wall defects in the 2D spin system have only a fixed energy cost independent of system size L rather than the usual linear scaling L; however their entropy (S) still diverges with system size, giving a free energy $E - TS$ (E = enthalpy and T = absolute temperature) that is infinitely negative relative to the ground state, hence a finite density of domain wall defects at any nonzero temperature. The spin-orbit and magnetoelastic couplings imply that each defect in the effective-spin texture also produces a change in the local orbitals and local distortions of the crystal.

Now consider the effect of applying a magnetic field. The magnetoelastic coupling requires us to go beyond the spin manifold and include ionic positions. Similarly, the magnetic field will also couple differently to the orbital and spin characters of the effective spins and thus modify the orbitals, again coupling to the crystal structure. Within a model such as the one represented by H_c, the degeneracy of the manifold containing each state implies that even infinitesimally small magnetic fields have a singular and large effect on the free energy landscape. The manifolds with the lowest free energy (and large defect density relative to the ground state) will be split and rearranged so as to favor states with uniform and nonzero magnetization, with those states forming a complicated landscape of free energy barriers reminiscent of spin glasses. If taken literally, this argument would also suggest that the dynamics of the crystal formation can proceed more easily through states with many proliferated defects, and thus are less likely to lead to any state with a particular long-range pattern of distortions; this would be consistent with the apparent experimental result that Ca_2RuO_4 and Sr_2IrO_4 have less crystalline distortions (such as octahedra tilts) upon field altering, but we caution that this part of the theoretical argument may not generalize beyond the fine-tuned H_c Hamiltonian.

The full Hamiltonian has many strongly competing terms beyond any single fine-tuned-frustration term such as H_c. When the full Hamiltonian still results in magnetic frustration, as indeed is observed in all three non-altered compounds discussed above, the frustrated system no longer has degenerate manifolds, but instead shows a glass-like landscape of states, some of which lie nearby in energy but have drastically different spin configurations. As the crystal forms and samples this free energy landscape, even small magnetic field magnitudes can have a large effect on the dynamics of the relaxation of crystal distortions and electronic structure [1].

The change in orbital distortions is observable not only in the structural and magnetic properties, but also in electron hopping and resistivity. The direct observation of this connection was recently argued for Ca_2RuO_4, and it is reasonable to conjecture that Sr_2IrO_4 will exhibit an analogous but stronger effect due to its stronger spin-orbit interaction. Note also that the H_c toy model can be applicable to Ca_2RuO_4 and Sr_2IrO_4, but not $Ba_4Ir_3O_{10}$; however, the immense frustration seen in non-altered $Ba_4Ir_3O_{10}$ implies that an analogous mechanism is, nevertheless, likely at play as well. These issues will require further microscopic analysis [1].

In conclusion, all results have clearly demonstrated that the field-altering technology is extraordinarily effective for generating new quantum states in correlated and spin-orbit-coupled materials. It is particularly astonishing that all this is achieved via an applied magnetic field no stronger than 0.06 Tesla applied during materials growth. With stronger magnetic fields, this technology should overcome even stronger energy barriers to altered quantum states that cannot be otherwise produced.

Further Reading

- Brian Pamplin, 2nd Ed. *Crystal Growth*. Pergamon (1980)
- S.M. Koohpayeh, D. Fort, and J.S. Abell. The optical floating zone technique: A review of experimental procedures with special reference to oxides. *Progress in Crystal Growth and Characterization of Materials* 54, 121 (2008)

References

1. Cao, G., Zhao, H., Hu, B., Pellatz, N., Reznik, D., Schlottmann, P., Kimchi, I. Quest for quantum states via field-altering technology, *npj Quantum Materials* 5, 83 (2020)
2. Pamplin, B.R., *Crystal Growth*, 2nd Ed. (Oxford: Pergamon, 1980)
3. Koohpayeh, S.M., Fort, D., Abell, J.S. The optical floating zone technique: a review of experimental procedures with special reference to oxides, *Progress in Crystal Growth and Characterization of Materials* 54, 121 (2008)
4. Tachibana, M. *Beginner's guide to flux crystal growth.* (Berlin: Springer-Verlag, 2017)

5. Macquart, R.B., Smith, M.D., zur Loye, H.-C. Crystal growth and single-crystal structures of *RE*RhO$_3$ (*RE* = La, Pr, Nd, Sm, Eu, Tb) orthorhodites from a K$_2$CO$_3$ flux, *Crystal Growth and Design* 6, 1361 (2006)

6. Wanklyn, B.M. Effects of modifying starting compositions for flux growth, *J. Cryst. Growth* 43, 336 (1978)

7. Ye, F., Chi, S., Chakoumakos, B.C., Fernandez-Baca, J.A., Qi, T., Cao, G. Magnetic and crystal structures of Sr$_2$IrO$_4$: a neutron diffraction study, *Phys. Rev. B* 87, 140406(R) (2013)

8. Ye, F., Chi, S., Cao, H., Chakoumakos, B., Fernandez-Baca, J.A., Custelcean, R., Qi, T., Korneta, O.B., Cao, G. Direct evidence of a zigzag spin-chain structure in the honeycomb lattice: a neutron and X-ray diffraction investigation of single-crystal Na$_2$IrO$_3$, *Phys. Rev. B* 85, 180403(R) (2012)

9. Steiner, U.E., Ulrich, T. Magnetic field effects in chemical kinetics and related phenomena, *Chem. Rev.* 89, 51 (1989)

10. Dold, P., Croll, A., Benz, K.W. Floating-zone growth of silicon in magnetic fields I. Weak static axial fields, *Journal of Crystal Growth* 183, 545 (1998)

11. Dold, P., Benz, K.W. Modification of fluid flow and heat transport in vertical Bridgman configurations by rotating magnetic fields, *Cryst. Res. Technol.* 32, 51 (1997)

12. Series, R.W., Hurle, D.T.J. The use of magnetic fields in semiconductor crystal growth, *Journal of Crystal Growth* 113, 305 (1991)

13. Wakayama, N.I. Effects of a strong magnetic field on protein crystal growth, *Crystal Growth and Design* 3, 1, 17 (2003)

14. Nussinov, Z., van den Brink, J. Compass models: theory and physical motivations, *Rev. Mod. Phys.* 87, 1 (2015)

Appendix
Synopses of Selected Experimental Techniques

A. Angle-Resolved Photoemission Spectroscopy (ARPES)

ARPES, a photoelectric effect, measures the distribution of energy and momentum of electrons ejected by photons from a material of interest [1]. The ejected electrons carry information on the electronic band structure and interactions in the material. The penetration depth of light ranges from a few nanometers to a few hundreds of nanometers, depending on the photon energy. A high-quality sample surface is thus crucial for ARPES measurements. ARPES has been widely exploited for studies of correlated electron systems since the discovery of the high-Tc cuprates in 1986, and in recent years, for studies of topological materials.

B. Muon Spin Resonance or Rotation (μSR)

μSR probes the influence of the local environment on the spin-polarized muons that are implanted in a material under study. The muon is a spin-1/2 particle and has a lifetime of 2.2×10^{-6} s. Its decay emits a positron from the material, and it is the emitted positron that yields the information on the interaction between the muon and its local environment in the material [2]. Single-crystal materials are not necessary because μSR data carry no spatial information.

C. Neutron Diffraction and Inelastic Neutron Scattering

Neutron diffraction entails *elastic* neutron scattering. Similar to X-ray diffraction, it is an essential tool commonly utilized to determine crystal structures. Neutron diffraction has a large penetration depth; therefore it yields more comprehensive structural details than does X-ray diffraction. Large single-crystal samples are essential for neutron diffraction. This is particularly true for materials containing heavy elements, such as iridates, because of the large neutron absorption cross-section, a measure of the likelihood of neutron capture. In contrast, a much smaller sample (e.g., ~70 μm) is more desirable in the case of single-crystal X-ray diffraction in order to reduce the X-ray absorption.

Inelastic neutron scattering is a probe exploited to study atomic motions (e.g., phonon modes), as well as magnetic crystal field excitations. The materials properties are obtained through measuring changes in energy and momentum of a scattered neutron due to the inelastic collision of the neutron with the sample. This technique generally requires large single crystals for good results.

D. Raman Scattering

Raman scattering, which involves inelastic scattering of photons, probes changes in photon energy to gain information on the properties of the material from which the photons are scattered (elastic scattering of photons is called Rayleigh scattering). The incident photons can excite vibrational modes of the material by transferring a certain amount of energy to the material. Scattering that causes decreased photon energy is called Stokes Raman scattering; scattering that causes increased photon energy is called anti-Stokes Raman scattering.

E. Resonant Inelastic X-Ray Scattering (RIXS)

RIXS probes the valence electrons, electronic correlations, and collective modes of materials. It is a photon-in/photon-out spectroscopy, thus the differences in energy and momentum between the incident and scattered photons directly correspond to the energy and momentum transferred from the incident photons to electrons of materials. In particular, magnetic excitations of iridates, which are often in a range of 10 meV to 100 meV, have been extensively studied via RIXS [3] because the Ir L_3-edge falls in a favorable energy range of X-rays ($\sim$ 11 keV). On the other hand, RIXS is rarely used, if at all, for studies of ruthenates and other $4d$-transition metal materials because the Ru L_3-edge lies inconveniently in the intermediate energy of X-rays ($\sim$2.8 keV for Ru, 2.5–3.5 keV for $4d$-elements) due chiefly to the lack of suitable optical schemes [4].

F. Second-Harmonic Generation (SHG)

SHG or frequency doubling is a phenomenon that can occur in nonlinear crystals or crystals without inversion symmetry, which can have a strong second-order nonlinear susceptibility. It was first observed in 1961 [5]. In essence, two photons with the same frequency interact with a non-linear crystal to produce a new photon with twice the frequency of the initial two photons in the nonlinear optical process. This technique has recently provided a powerful tool for studies of many spin-orbit-coupled materials, such as Sr_2IrO_4 [6].

G. X-Ray Absorption Spectroscopy (XAS)

XAS probes the local or electronic structure of materials using a tunable X-ray photon source with energies ranging from 0.1 keV to 100 keV, which can excite core electrons. The XAS edges (namely, K-, L-, and M-edges) of core electrons correspond to the principle atomic quantum numbers n = 1, 2, and 3, respectively. It is a widely used technique for studies of gases, liquids, and solids.

H. X-Ray Magnetic Circular Dichroism (XMCD)

XMCD probes magnetic properties, such as spin and orbital magnetic moments, of materials via measurements of X-ray absorption spectra (XAS) for both right- and left-circularly polarized light in an applied magnetic field. The difference between the two spectra yields the information on the magnetic properties of a material of interest. Absorption spectra for transition metal oxides are measured at the L-edges of transition elements, which are particularly sensitive to the magnetic properties of transition metal oxides [7]. Unlike neutron scattering or diffraction, XMCD does not require large single-crystal samples. In recent years it has been widely used to investigate the magnetic properties of iridates [8–10].

References

1. Hüfner, S. (Ed.) *Very high resolution photoelectron spectroscopy.* (Berlin: Springer-Verlag, 2007)
2. Blundell, S.J. Spin-polarized muons in condensed matter physics, *Contemp. Phys.* 40, 175 (1999)
3. Gretarsson, H., Sung, N., Porras, J., Bertinshaw, J., Dietl, C., Bruin, J.A., Bangura, A., Kim, Y., Dinnebier, R., Kim, J., Al-Zein, A., Moretti Sala, M., Krisch, M., Le Tacon, M., Keimer, B., Kim, B. Persistent Paramagnons Deep in the Metallic Phase of $Sr_{2-x}La_xIrO_4$, *Phys. Rev. Lett.* 117, 107001 (2016)
4. Gretarsson, H., Ketenoglu, D., Harder, M., Mayer, S., Dill, F.-U., Spiwek, M., Schulte-Schrepping, H., Tischer, M., Wille, H.-C., Keimer, B., Yavaş, H. IRIXS: a resonant inelastic X-ray scattering instrument dedicated to X-rays in the intermediate energy range, *J. Synchrotron Rad.* 27, 538 (2020)
5. Franken, P.A., Hill, A.E., Peters, C.W., Weinreich, G. Generation of optical harmonics, *Phys. Rev. Lett.* 7, 118 (1961)
6. Zhao, L., Torchinsky, D.H., Chu, H., Ivanov, V., Lifshitz, R., Flint, R., Qi, T., Cao, G., Hsieh, D. Evidence of an odd-parity hidden order in a spin-orbit coupled correlated iridate, *Nature Physics.* 12, 32 (2016)
7. Laguna-Marco, M.A., Haskel, D., Souza-Neto, N., Lang, J.C., Krishnamurthy, V.V., Chikara, S., Cao, G., van Veenendaal, M. Orbital magnetism and spin-orbit effects in the electronic structure of $BaIrO_3$, *Phys. Rev. Lett.* 105, 216407 (2010)
8. Haskel, D., Fabbris, G., Zhernenkov, M., Kong, P.P., Jin, C.Q., Cao, G., van Veenendaal, M. Pressure-tuning of spin-orbit coupled ground state in Sr_2IrO_4, *Phys. Rev. Lett.* 109, 027204 (2012)
9. Takayama, T., Kato, A., Dinnebier, R., Nuss, J., Kono, H., Veiga, L.S.I., Fabbris, G., Haskel, D., Takagi, H. Hyperhoneycomb iridate β-Li_2IrO_3 as a platform for Kitaev magnetism, *Phys. Rev. Lett.* 114, 077202 (2015)
10. Haskel, D., Fabbris, G., Kim, J.H., Veiga, L.S.I., Mardegan, J.R.L., Escanhoela, C.A. Jr., Chikara, S., Struzhkin, V., Senthil, T., Kim, B.J., Cao, G., Kim, J.W. Possible quantum paramagnetism in compressed Sr_2IrO_4, *Phys. Rev. Lett.* 124, 067201 (2020)

Subject Index

Compound Index